Hydraulic Design Tables for Pipelines, Sewers and Drainage Lines

(Relating Head Loss v/s Power Loss)

Based on Hazen Williams and Modified
Hazen Williams Formulae

Hydraulic Design Tables for Pipelines, Sewers and Drainage Lines

(Relating Head Loss v/s Power Loss)

Based on Hazen Williams and Modified Hazen Williams Formulae

J.N. Joshi

Former Professor of Environmental Engineering
L.D. College of Engineering, Ahmedabad
Gujarat, India

PUBLISHING FOR ONE WORLD

NEW AGE INTERNATIONAL (P) LIMITED, PUBLISHERS

(formerly Wiley Eastern Limited)

New Delhi • Bangalore • Chennai • Guwahati • Hyderabad
Jalandhar • Kolkata • Lucknow • Mumbai • Ranchi

Visit us at **www.newagepublishers.com**

Published by New Age International (P) Ltd., Publishers
First Edition: 2006
Reprint: 2009

Branches:

- 36, Malikarjuna Temple Street, Opp. ICWA, Basavanagudi, **Bangalore**. ✆ (080) 26677815
- 26, Damodaran Street, T. Nagar, **Chennai**. ✆ (044) 24353401
- Hemsen Complex, Mohd. Shah Road, Paltan Bazar, Near Starline Hotel, **Guwahati**. ✆ (0361) 2543669
- No. 105, 1st Floor, Madhiray Kaveri Tower, 3-2-19, Azam Jahi Road, Nimboliadda, **Hyderabad**. ✆ (040) 24652456
- RDB Chambers (Formerly Lotus Cinema) 106A, Ist Floor, S.N. Banerjee Road, **Kolkata**. ✆ (033) 22275247
- 16-A, Jopling Road, **Lucknow**. ✆ (0522) 2209578
- 142C, Victor House, Ground Floor, N.M. Joshi Marg, Lower Parel, **Mumbai**. ✆ (022) 24927869
- 22, Golden House, Daryaganj, **New Delhi**. ✆ (011) 23262370, 23262368

ISBN(10): 81-224-1867-8

ISBN(13) : 978-81-224-1867-5

Rs. 195.00

C-09-06-3732

Printed in India at Glorious Printers, Delhi.
Typeset at Shubham Graphics, Delhi.

PUBLISHING FOR ONE WORLD
NEW AGE INTERNATIONAL (P) LIMITED, PUBLISHERS
4835/24, Ansari Road, Daryaganj, New Delhi-110002
Visit us at **www.newagepublishers.com**

Foreword

Professor J N Joshi has evolved the Hydraulic Design Tables existing in literature. His pro-academic attitude has incorporated pragmatic, on the field practices. Professor Joshi is an illustration of step by step progress. He started his academic career as an Assistant Lecturer and culminated as Professor of Environmental Engineering , through Public Service Commission selections. He received a Merit Award for ranking first in his M E (Environmental).

The booklet is useful to not only students in their project work, but also to practicing engineers and consultants, the focus being overall economy. It includes tables for design of water supply net works and waste water systems.

I congratulate Professor J N Joshi for this compilation in the pursuit of excellence.

P.P. Oza
B.E (Hons.), Ph.D.
Former Advisor (Scitech),
Government of Gujarat.
Former Professor of Environmental Engineering ,
L.D. College of Engineering, Ahmedabad

Preface

It gives me great pleasure to present this booklet entitled 'Most Useful Hydraulic Design Tables for Pipe Lines, Sewers and Drainage lines' in the hands of my friends of engineering fraternity of all the disciplines working in the field of hydraulics in active service or practicing engineers and consultants.

First booklet in this subject was published by the eminent Engineer Late Shri D B Bhatt long back in 1969. It is a standard reference as on today. Then came a second booklet with excellent printing and encompassing new and novel features by an experienced Engineer Shri A G Desai in 2001.But still I felt that there was a need for the values of velocities that are mostly used in real field and to avoid unnecessary small values of units and digits after decimal points.

The purpose of this type of booklet , according to me should be to focus attention of designing engineers towards *overall economy* of the water supply Schemes. It is evident that head lost in overcoming friction is a mere waste of energy and ultimately money. Therefore a new feature is added in this booklet which relates *Power* to over come friction to transmit certain quantity of water through a distance of one km.

Hope this endeavour of mine will be useful to practicing engineers and consultants as well in the field of water supply and sanitary engineering and industrial utilities.

The focus on power consumption will lead to awareness of energy conservation as a '*unit saved is a unit produced*'. Looking to this point in view I feel that an exposure of this booklet to Engineering Students will go a long way in future energy conservation drive.

Last but not the least I am grateful to my teacher, friend, philosopher and guide Dr P P Oza for being my constant source of inspiration and support and for his kind gesture in writing Foreword of this booklet.

Ahmedabad

J N Joshi

LIST OF ABBRIVIATIONS USED

Sqm	:	square meter
E-xx	:	10 to the power (-xx)
V	:	velocity of flow, m/s
kL/d	:	kilo liter per day or : cubic meter per day
kL/h	:	kilo liter per hour or : cubic meter per hour
kL/min	:	kilo liter per minute or : cubic meter per minute
ML/d	:	million liter per day
H_{fc}, m/km	:	head loss in meter per kilo meter as per conventional Hazen Williams Formulae
H_{fm}	:	head loss in meter per kilometer as per modified Hazen Williams Formulae
kW/km	:	kilo watt of power wasted due to pipe friction per kilo meter as per conventional Hazen Williams Formulae
C	:	Hazen Williams friction factor
f_q	:	multiplying factor for flow rate
f_{hc}	:	multiplying factor for head loss as per conventional formulae
f_{hm}	:	multiplying factor for head loss as per modified formulae
n	:	Mannings constant for roughness of surface in contact with water
nr	:	ratio of n values of partially flowing to fully flowing pipe section
dr	:	ratio of depths of flow partially flowing to that of fully flowing pipe
pr	:	ratio of wetted perimeter of partially flowing to that of fully flowing pipe
ar	:	ratio of hydraulic radius of partially flowing to that of fully flowing pipe
vr	:	ratio of velocity of partially flowing to that of fully flowing pipe
qr	:	ratio of flow rates of partially flowing to that of fully flowing pipe

FORMULAE USED

SECTION-I
DISCHARGE AND HEAD AND POWER LOSS IN PIPES FLOWING UNDER PRESSURE

Q_M = FLOW, ML/d

$H_f = 2.237 \times 10^{11} \times Q_M^{1.852} \times D^{-4.87}$

$H_{fm} = 8.4543 \times 10^{10} \times Q_M^{1.81} \times D^{-4.81}$

D = Diameter, mm

Q_d = FLOW, cum/d

$H_{fC} = 6.217 \times 10^5 \times Q_d^{\ 1.852} \times D^{-4.87}$

$H_{fm} = 3.141 \times 10^5 \times Q_d^{1.81} \times D^{-4.81}$

H_{fC} = Head lost in friction, m / km as per H-W formulae

H_{fm} = Head lost in friction as per Modified H-W formulae, m / km

Q_h = FLOW, cum/h, Q_{min}= cum/min

V = Velocity, m/s

$Q_M = 6.786 \times 10^{-5} \times V \times D^2$, ML/d

$Q_d = 6.786 \times 10^{-2} \times V \times D^2$, cum/d

$Q_h = 2.828 \times 10^{-3} \times V \times D^2$, cum/h

$Q_{min} = 4.712 \times 10^{-5} \times V \times D^2$, cum/min

kW/km = Power Lost per km in over coming Frictional Head Loss

kW/km = 0.2778 * Qmin * H_{fc}

kW/km = 4.63 * 10^{-3} * Q_h * H_{fc}

kw/km = 1.929 * 10^{-4} * Q_d * H_{fc}

kw/km = 0.1929 * Q_M * H_{fc}

FORMULAE USED

SECTION-II

DISCHARGE & GRADIENT FOR PIPES FLOWING FULL UNDER GRAVITY CONDITIONS

$QM = 67.86 * V * D^2$ $\quad Q_M$ = Flow in ML/d, D = Dia. in m

$Q_{min} = Q_M/1.44$ $\quad Q_{min}$= Flow in cum/min

Gradient : 1:L

$L = 0.1575 * D^{4/3}/(VN)^2$ $\quad$ L = Length in m

V = Velocity in m/s

N = Mannings constant

SECTION-III

PARAMETERS OF PARTIALLY FLOWING PIPES

$\theta = 2 * \cos^{-1}(1 - 2d/D)$

$a/A = (1/2\pi) * (\theta - \sin(\theta) = 0.159 * (\theta - \sin(\theta))$

$r/R = (1 - \sin(\theta)/\theta)$

$r/R = (1 - \sin(\theta)/\theta)$

$v/V = (N/n) * (r/R)^{2/3}$

$q/Q = (N/n) * (a/A) * (r/R)^{2/3}$ Where,

θ = Central Angle in Radians

(1 Radian = 180/Π = 57.296 0)

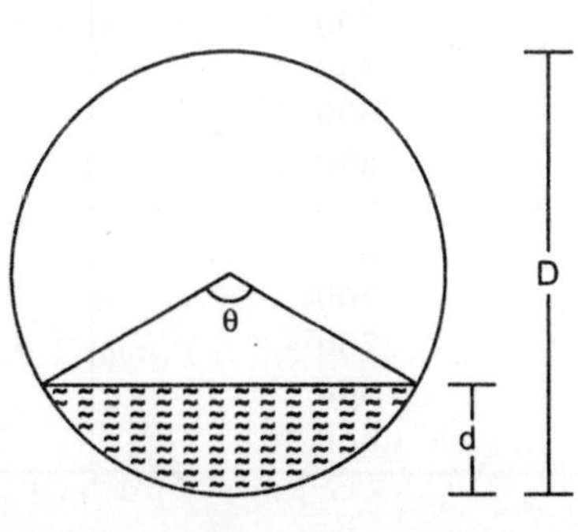

a, r, n, v, and q refers to area, hydraulic radius, Mannings constant velocity and discharge pertaining to partially flowing section and capital letters pertain to fully flowing section respectively.

List of Diameters and their Areas

Diameter, mm	*Area, sqm*	*Diameter, mm*	*Area, sqm*
15	0.000177	900	0.6362
20	0.000314	950	0.7088
25	0.000491	1000	0.7854
32	0.000804	1050	0.8659
40	0.001257	1100	0.9503
50	0.001964	1150	1.0387
65	0.003318	1200	1.131
80	0.005027	1250	1.2272
100	0.007854	1300	1.3273
125	0.012272	1400	1.5394
140	0.015394	1500	1.7672
150	0.017672	1550	1.8869
170	0.022698	1600	2.0106
190	0.028353	1650	2.1383
200	0.031416	1700	2.2698
225	0.039761	1750	2.4053
250	0.049088	1800	2.5447
270	0.057256	1850	2.688
300	0.070686	1900	2.8353
350	0.096212	2000	3.1416
400	0.125664	2100	3.4636
450	0.159044	2200	3.8013
500	0.19635	2300	4.1548
550	0.237584	2400	4.5239
600	0.282744	2500	4.9088
650	0.331832	2600	5.3093
700	0.384846	2700	5.7256
750	0.441788	2800	6.1575
800	0.502656	3000	7.0686
850	0.567452	3100	7.5477

Contents

SECTION I

Pipes Flowing Under Pressure
Tables for Discharge and Loss of Head and Power

TABLE: HW 01 DISCHARGE AND LOSS OF HEAD AND POWER IN PIPES
(Hazen-Williams Formula, C=100)
(Modified Hazen-Williams Formula, CR=1)

Diameter of pipe 15 mm **Area 1.7672E-04 sqm**

Velocity V, m/s	*Flow kL/d*	*Flow kL/h*	*Head Loss Hfc,m/km*	*Head Loss Hfm,m/km*	*kW Wasted/km 60% efficiency*
1	2	3	4	5	6
0.4	6.1	0.25	32	17.7	0.04
0.45	6.9	0.29	42.2	23.2	0.06
0.5	**7.6**	**0.32**	**50.6**	**27.7**	**0.07**
0.55	8.4	0.35	59.8	32.6	0.1
0.6	9.2	0.38	69.6	37.8	0.12
0.65	9.9	0.41	80.1	43.4	0.15
0.7	10.7	0.45	95.2	51.4	0.2
0.75	11.5	0.48	107.3	57.7	0.24
0.8	12.2	0.51	120	64.4	0.28
0.85	13	0.54	133.4	71.4	0.33
0.9	13.7	0.57	147.4	78.8	0.39
0.95	14.5	0.6	162.1	86.4	0.45
1	**15.3**	**0.64**	**182.7**	**97.1**	**0.54**
1.05	16	0.67	198.9	105.5	0.62
1.1	16.8	0.7	215.7	114.3	0.7
1.15	17.6	0.73	233.1	123.3	0.79
1.2	18.3	0.76	251.2	132.6	0.88
1.25	19.1	0.8	276.2	145.5	1.02
1.3	19.8	0.83	295.7	155.5	1.14
1.35	20.6	0.86	315.8	165.8	1.26
1.4	21.4	0.89	336.5	176.5	1.39
1.45	22.1	0.92	357.8	187.4	1.52
1.5	**22.9**	**0.95**	**379.7**	**198.6**	**1.67**
1.55	23.7	0.99	409.9	214	1.88
1.6	24.4	1.02	433.2	225.8	2.05
1.65	25.2	1.05	457.1	238	2.22
1.7	**26**	**1.08**	**481.5**	**250.5**	**2.41**
1.75	26.7	1.11	506.6	263.2	2.6
1.8	27.5	1.15	540.9	280.6	2.88
1.85	28.2	1.18	567.4	294	3.1
1.9	29	1.21	594.4	307.7	3.33
1.95	29.8	1.24	622	321.6	3.57
2	**30.5**	**1.27**	**650.1**	**335.8**	**3.82**
2.05	31.3	1.3	678.8	350.3	4.09
2.1	32.1	1.34	718	370.1	4.45
2.15	32.8	1.37	748.1	385.2	4.75
2.2	33.6	1.4	778.7	400.6	5.05
2.25	34.4	1.43	809.9	416.3	5.36
2.3	35.1	1.46	841.6	432.2	5.69
2.35	35.9	1.5	884.8	453.9	6.14
2.4	**36.6**	**1.53**	**917.9**	**470.47**	**6.5**

C	80	90	100	110	120	130	140	145	150
Fq	.80	.90	1.00	1.10	1.20	1.30	1.40	1.45	1.50
Fhc	1.511	1.215	1.000	0.838	0.713	0.615	0.536	0.503	0.472
Fhm	1.498	1.210	1.000	0.842	0.719	0.622	0.544	0.510	0.480

Multiplying factors fq, fhc and fhm for dischange, head loss and modified head loss respectively wrt C values.

TABLE: HW 02 DISCHARGE AND LOSS OF HEAD AND POWER IN PIPES (Hazen-Williams Formula, C=100) (Modified Hazen-Williams Formula, CR=1)

Diameter of pipe 20 mm **Area 3.1416E-04 sqm**

Velocity V, m/s	*Flow kL/d*	*Flow kL/h*	*Head Loss Hfc,m/km*	*Head Loss Hfm,m/km*	*kW Wasted/km 60% efficiency*
1	2	3	4	5	6
0.4	10.9	0.45	23.4	12.9	0.05
0.45	12.2	0.51	29.6	16.1	0.07
0.5	**13.6**	**0.57**	**36.3**	**19.7**	**0.1**
0.55	14.9	0.62	42.4	23	0.12
0.6	16.3	0.68	50.4	27.2	0.16
0.65	17.6	0.73	57.4	30.9	0.19
0.7	19	0.79	66.5	35.6	0.24
0.75	20.4	0.85	76.1	40.7	0.3
0.8	21.7	0.9	84.6	45.1	0.35
0.85	23.1	0.96	95.4	50.7	0.42
0.9	24.4	1.02	106.7	56.6	0.5
0.95	25.8	1.08	118.6	62.8	0.59
1	**27.1**	**1.13**	**129**	**68.1**	**0.67**
1.05	28.5	1.19	142	74.8	0.78
1.1	29.9	1.25	155.5	81.8	0.9
1.15	31.2	1.3	167.2	87.8	1.01
1.2	32.6	1.36	181.8	95.3	1.14
1.25	33.9	1.41	194.4	101.7	1.27
1.3	35.3	1.47	210	109.7	1.43
1.35	36.6	1.53	226.1	117.9	1.6
1.4	38	1.58	240	125	1.76
1.45	39.4	1.64	257.1	133.7	1.95
1.5	**40.7**	**1.7**	**274.8**	**142.7**	**2.16**
1.55	42.1	1.75	290	150.4	2.35
1.6	43.4	1.81	308.7	159.8	2.59
1.65	44.8	1.87	327.9	169.6	2.84
1.7	46.1	1.92	344.3	177.9	3.06
1.75	47.5	1.98	364.5	188	3.34
1.8	48.9	2.04	385.2	198.5	3.64
1.85	50.2	2.09	402.9	207.4	3.9
1.9	51.6	2.15	424.6	218.3	4.23
1.95	52.9	2.2	443.1	227.5	4.51
2	**54.3**	**2.26**	**465.7**	**238.9**	**4.87**
2.05	55.6	2.32	488.8	250.5	5.25
2.1	57	2.38	512.5	262.4	5.65
2.15	58.4	2.43	532.6	272.4	5.99
2.2	59.7	2.49	557.2	284.7	6.42
2.25	61.1	2.55	582.4	297.3	6.88
2.3	62.4	2.6	603.7	307.9	7.27
2.35	63.8	2.66	629.8	320.9	7.76
2.4	**65.1**	**2.71**	**651.9**	**331.9**	**8.18**

C	80	90	100	110	120	130	140	145	150
Fq	.80	.90	1.00	1.10	1.20	1.30	1.40	1.45	1.50
Fhc	1.511	1.215	1.000	0.838	0.713	0.615	0.536	0.503	0.472
Fhm	1.498	1.210	1.000	0.842	0.719	0.622	0.544	0.510	0.480

Multiplying factors fq, fhc and fhm for dischange, head loss and modified head loss respectively wrt C values.

TABLE: HW 03 DISCHARGE AND LOSS OF HEAD AND POWER IN PIPES
(Hazen-Williams Formula, C=100)
(Modified Hazen-Williams Formula, CR=1)

Diameter of pipe 25 mm **Area 4.9088E-04 sqm**

Velocity V, m/s	*Flow kL/d*	*Flow kL/h*	*Head Loss Hfc,m/km*	*Head Loss Hfm,m/km*	*kW Wasted/km 60% efficiency*
1	2	3	4	5	6
0.4	17	0.71	18.4	10	0.06
0.45	19.1	0.8	23	12.5	0.09
0.5	**21.2**	**0.88**	**27.4**	**14.8**	**0.11**
0.55	23.3	0.97	32.8	17.7	0.15
0.6	25.4	1.06	38.7	20.7	0.19
0.65	27.6	1.15	45	24	0.24
0.7	29.7	1.24	51.7	27.6	0.3
0.75	31.8	1.33	58.8	31.3	0.36
0.8	33.9	1.41	65.6	34.8	0.43
0.85	36.1	1.5	73.5	38.9	0.51
0.9	38.2	1.59	81.9	43.2	0.6
0.95	40.3	1.68	90.7	47.7	0.71
1	**42.4**	**1.77**	**99.9**	**52.5**	**0.82**
1.05	44.5	1.85	108.4	56.8	0.93
1.1	46.7	1.95	119.5	62.5	1.08
1.15	48.8	2.03	128.8	67.3	1.21
1.2	50.9	2.12	139.5	72.7	1.37
1.25	53	2.21	150.7	78.4	1.54
1.3	55.1	2.3	162.3	84.3	1.73
1.35	57.3	2.39	174.2	90.4	1.93
1.4	59.4	2.48	186.6	96.6	2.14
1.45	61.5	2.56	197.9	102.3	2.35
1.5	**63.6**	**2.65**	**211**	**108.9**	**2.59**
1.55	65.7	2.74	224.4	115.7	2.85
1.6	67.9	2.83	238.3	122.7	3.12
1.65	70	2.92	252.5	129.9	3.41
1.7	72.1	3	265.4	136.4	3.69
1.75	74.2	3.09	280.4	143.9	4.01
1.8	76.3	3.18	295.7	151.5	4.35
1.85	78.5	3.27	311.4	159.4	4.71
1.9	80.6	3.36	327.4	167.4	5.09
1.95	82.7	3.45	343.9	175.6	5.49
2	**84.8**	**3.53**	**358.8**	**183.1**	**5.86**
2.05	86.9	3.62	375.9	191.6	6.3
2.1	89.1	3.71	393.4	200.3	6.76
2.15	91.2	3.8	411.2	209.2	7.23
2.2	93.3	3.89	429.5	218.3	7.74
2.25	95.4	3.98	448	227.5	8.26
2.3	97.5	4.06	464.9	235.8	8.74
2.35	99.7	4.15	484.1	245.4	9.3
2.4	**101.8**	**4.24**	**503.8**	**255.1**	**9.89**

C	80	90	100	110	120	130	140	145	150
Fq	.80	.90	1.00	1.10	1.20	1.30	1.40	1.45	1.50
Fhc	1.511	1.215	1.000	0.838	0.713	0.615	0.536	0.503	0.472
Fhm	1.498	1.210	1.000	0.842	0.719	0.622	0.544	0.510	0.480

Multiplying factors fq, fhc and fhm for dischange, head loss and modified head loss respectively wrt C values.

TABLE: HW 04 DISCHARGE AND LOSS OF HEAD AND POWER IN PIPES
(Hazen-Williams Formula, C=100)
(Modified Hazen-Williams Formula, CR=1)

Diameter of pipe 32 mm **Area 8.0425E-04 sqm**

Velocity V, m/s	*Flow kL/d*	*Flow kL/h*	*Head Loss Hfc,m/km*	*Head Loss Hfm,m/km*	*kW Wasted/km 60% efficiency*
1	2	3	4	5	6
0.4	27.8	1.16	13.7	7.4	0.07
0.45	31.3	1.3	17	9.2	0.1
0.5	**34.7**	**1.45**	**20.8**	**11.2**	**0.14**
0.55	38.2	1.59	24.6	13.2	0.18
0.6	41.7	1.74	29.1	15.5	0.23
0.65	45.2	1.88	33.6	17.9	0.29
0.7	48.6	2.03	38.7	20.5	0.36
0.75	52.1	2.17	43.8	23.1	0.44
0.8	55.6	2.32	49.6	26.1	0.53
0.85	59.1	2.46	55.2	29	0.63
0.9	62.5	2.6	61.2	32.1	0.74
0.95	66	2.75	67.9	35.5	0.86
1	**69.5**	**2.9**	**74.9**	**39.1**	**1.01**
1.05	73	3.04	81.8	42.6	1.15
1.1	76.4	3.18	88.9	46.2	1.31
1.15	79.9	3.33	96.8	50.2	1.49
1.2	83.4	3.48	105	54.4	1.69
1.25	86.9	3.62	113	58.4	1.89
1.3	90.3	3.76	121.2	62.6	2.11
1.35	93.8	3.91	130.3	67.2	2.36
1.4	97.3	4.05	139.1	71.6	2.61
1.45	100.8	4.2	148.8	76.5	2.89
1.5	**104.2**	**4.34**	**158.1**	**81.2**	**3.18**
1.55	107.7	4.49	168.3	86.3	3.5
1.6	111.2	4.63	178.2	91.2	3.82
1.65	114.7	4.78	189	96.7	4.18
1.7	118.1	4.92	199.4	101.8	4.54
1.75	121.6	5.07	210.8	107.5	4.95
1.8	125.1	5.21	221.7	113	5.35
1.85	128.6	5.36	233.7	118.9	5.8
1.9	132	5.5	245.1	124.6	6.24
1.95	135.5	5.65	257.6	130.8	6.74
2	**139**	**5.79**	**269.6**	**136.7**	**7.23**
2.05	142.5	5.94	282.7	143.2	7.77
2.1	145.9	6.08	295.1	149.4	8.31
2.15	149.4	6.23	308.8	156.1	8.91
2.2	152.9	6.37	321.7	162.5	9.49
2.25	156.3	6.51	334.9	169.1	10.09
2.3	159.8	6.66	349.4	176.2	10.77
2.35	163.3	6.8	363.1	182.9	11.43
2.4	**166.8**	**6.95**	**378.1**	**190.3**	**12.17**

C	80	90	100	110	120	130	140	145	150
Fq	.80	.90	1.00	1.10	1.20	1.30	1.40	1.45	1.50
Fhc	1.511	1.215	1.000	0.838	0.713	0.615	0.536	0.503	0.472
Fhm	1.498	1.210	1.000	0.842	0.719	0.622	0.544	0.510	0.480

Multiplying factors fq, fhc and fhm for dischange, head loss and modified head loss respectively wrt C values.

TABLE: HW 05 DISCHARGE AND LOSS OF HEAD AND POWER IN PIPES

(Hazen-Williams Formula, C=100)

(Modified Hazen-Williams Formula, CR=1)

Diameter of pipe 40 mm **Area 1.2566E-03 sqm**

Velocity V, m/s	*Flow kL/d*	*Flow kL/h*	*Head Loss Hfc,m/km*	*Head Loss Hfm,m/km*	*kW Wasted/km 60% efficiency*
1	2	3	4	5	6
0.4	43.4	1.8	10.4	5.6	1.43
0.45	48.9	2	12.7	6.8	1.6
0.5	**54.3**	**2.3**	**16.5**	**8.8**	**1.79**
0.55	59.7	2.5	19.2	10.2	2
0.6	65.1	2.7	22.1	11.8	2.33
0.65	70.6	2.9	25.3	13.4	2.57
0.7	76	3.2	30.3	16	2.83
0.75	81.4	3.4	33.9	17.8	3.1
0.8	86.9	3.6	37.7	19.8	3.54
0.85	92.3	3.8	41.7	21.8	3.86
0.9	97.7	4.1	48	25	4.19
0.95	103.1	4.3	52.4	27.3	4.54
1	**108.6**	**4.5**	**57**	**29.6**	**5.1**
1.05	114	4.8	64.3	33.3	1.43
1.1	119.4	5	69.3	35.8	1.6
1.15	124.9	5.2	74.5	38.5	1.79
1.2	130.3	5.4	79.9	41.2	2
1.25	135.7	5.7	88.3	45.4	2.33
1.3	141.1	5.9	94.2	48.4	2.57
1.35	146.6	6.1	100.2	51.4	2.83
1.4	152	6.3	106.3	54.5	3.1
1.45	157.4	6.6	115.9	59.3	3.54
1.5	**162.9**	**6.8**	**122.5**	**62.5**	**3.86**
1.55	168.3	7	129.2	65.9	4.19
1.6	173.7	7.2	136.2	69.4	4.54
1.65	179.2	7.5	146.9	74.7	5.1
1.7	184.6	7.7	154.2	78.3	5.5
1.75	190	7.9	161.7	82	5.91
1.8	195.4	8.1	169.4	85.8	6.35
1.85	200.9	8.4	181.1	91.7	7.04
1.9	206.3	8.6	189.2	95.7	7.53
1.95	211.7	8.8	197.4	99.7	8.04
2	**217.2**	**9.1**	**210.1**	**106**	**8.85**
2.05	222.6	9.3	218.7	110.2	9.42
2.1	228	9.5	227.5	114.6	10.01
2.15	233.4	9.7	236.5	119	10.62
2.2	238.9	10	250.2	125.7	11.58
2.25	244.3	10.2	259.5	130.3	12.26
2.3	249.7	10.4	269	134.9	12.95
2.35	255.2	10.6	278.7	139.7	13.68
2.4	**260.6**	**10.9**	**293.5**	**146.9**	**14.81**

C	80	90	100	110	120	130	140	145	150
Fq	.80	.90	1.00	1.10	1.20	1.30	1.40	1.45	1.50
Fhc	1.511	1.215	1.000	0.838	0.713	0.615	0.536	0.503	0.472
Fhm	1.498	1.210	1.000	0.842	0.719	0.622	0.544	0.510	0.480

Multiplying factors fq, fhc and fhm for dischange, head loss and modified head loss respectively wrt C values.

TABLE: HW 06 DISCHARGE AND LOSS OF HEAD AND POWER IN PIPES
(Hazen-Williams Formula, C=100)
(Modified Hazen-Williams Formula, CR=1)

Diameter of pipe 50 mm **Area 1.9635E-03 sqm**

Velocity V, m/s	*Flow kL/d*	*Flow kL/h*	*Head Loss Hfc,m/km*	*Head Loss Hfm,m/km*	*kW Wasted/km 60% efficiency*
1	2	3	4	5	6
0.4	68	2.8	8	4.3	0.1
0.45	76	3.2	10.2	5.5	0.15
0.5	**85**	**3.5**	**12.1**	**6.4**	**0.2**
0.55	93	3.9	14.8	7.8	0.27
0.6	102	4.3	17.7	9.3	0.35
0.65	110	4.6	20	10.5	0.43
0.7	119	5	23.4	12.3	0.54
0.75	127	5.3	26	13.6	0.64
0.8	136	5.7	29.8	15.5	0.79
0.85	144	6	32.8	17	0.91
0.9	153	6.4	36.9	19.2	1.09
0.95	161	6.7	40.2	20.8	1.25
1	**170**	**7.1**	**44.8**	**23.1**	**1.47**
1.05	178	7.4	48.3	24.9	1.65
1.1	187	7.8	53.3	27.4	1.92
1.15	195	8.1	57.1	29.3	2.14
1.2	204	8.5	62.5	32	2.46
1.25	212	8.8	66.6	34.1	2.71
1.3	221	9.2	72.3	37	3.08
1.35	229	9.5	76.7	39.2	3.37
1.4	238	9.9	82.8	42.2	3.8
1.45	246	10.3	89.1	45.3	4.25
1.5	**254**	**10.6**	**94**	**47.8**	**4.61**
1.55	263	11	100.7	51.1	5.13
1.6	271	11.3	105.8	53.6	5.54
1.65	280	11.7	112.9	57.1	6.12
1.7	288	12	118.3	59.8	6.57
1.75	297	12.4	125.7	63.4	7.22
1.8	305	12.7	131.4	66.2	7.73
1.85	314	13.1	139.2	70.1	8.44
1.9	322	13.4	145.1	73	9
1.95	331	13.8	153.2	77	9.79
2	**339**	**14.1**	**159.5**	**80**	**10.41**
2.05	348	14.5	167.9	84.2	11.27
2.1	356	14.8	174.4	87.4	11.95
2.15	365	15.2	183.3	91.7	12.9
2.2	373	15.5	190	95	13.64
2.25	382	15.9	199.2	99.5	14.66
2.3	390	16.3	208.6	104.1	15.74
2.35	399	16.6	215.8	107.5	16.59
2.4	**407**	**17**	**225.5**	**112.3**	**17.75**

C	80	90	100	110	120	130	140	145	150
Fq	.80	.90	1.00	1.10	1.20	1.30	1.40	1.45	1.50
Fhc	1.511	1.215	1.000	0.838	0.713	0.615	0.536	0.503	0.472
Fhm	1.498	1.210	1.000	0.842	0.719	0.622	0.544	0.510	0.480

Multiplying factors fq, fhc and fhm for dischange, head loss and modified head loss respectively wrt C values.

TABLE: HW 07 DISCHARGE AND LOSS OF HEAD AND POWER IN PIPES
(Hazen-Williams Formula, C=100)
(Modified Hazen-Williams Formula, CR=1)

Diameter of pipe 65 mm **Area 3.3183E-03 sqm**

Velocity V, m/s	*Flow kL/d*	*Flow kL/h*	*Head Loss Hfc,m/km*	*Head Loss Hfm,m/km*	*kW Wasted/km 60% efficiency*
1	2	3	4	5	6
0.4	115	4.8	6	3.2	0.13
0.45	129	5.4	7.5	4	0.19
0.5	**143**	**6**	**9.1**	**4.8**	**0.25**
0.55	158	6.6	10.9	5.7	0.33
0.6	172	7.2	12.8	6.7	0.43
0.65	186	7.8	14.8	7.8	0.53
0.7	201	8.4	17	8.9	0.66
0.75	215	9	19.3	10.1	0.8
0.8	229	9.5	21.4	11.1	0.94
0.85	244	10.2	24.4	12.6	1.15
0.9	258	10.8	27.1	14	1.36
0.95	272	11.3	29.5	15.2	1.54
1	**287**	**12**	**33**	**16.9**	**1.83**
1.05	301	12.5	35.6	18.2	2.06
1.1	315	13.1	38.8	19.8	2.35
1.15	330	13.8	42.7	21.8	2.73
1.2	344	14.3	45.6	23.2	3.02
1.25	358	14.9	49.2	25	3.39
1.3	373	15.5	53	26.9	3.8
1.35	387	16.1	56.8	28.8	4.23
1.4	401	16.7	60.8	30.8	4.7
1.45	416	17.3	64.9	32.8	5.2
1.5	**430**	**17.9**	**69.1**	**34.9**	**5.73**
1.55	444	18.5	73.5	37	6.3
1.6	459	19.1	78	39.2	6.9
1.65	473	19.7	82.6	41.5	7.53
1.7	487	20.3	87.3	43.8	8.21
1.75	502	20.9	92.1	46.2	8.91
1.8	516	21.5	97.1	48.6	9.67
1.85	530	22.1	102.1	51.1	10.45
1.9	545	22.7	107.3	53.6	11.28
1.95	559	23.3	112.7	56.2	12.16
2	**573**	**23.9**	**118.1**	**58.9**	**13.07**
2.05	588	24.5	123.6	61.6	14.02
2.1	602	25.1	129.3	64.3	15.03
2.15	616	25.7	135.1	67.2	16.08
2.2	631	26.3	141	70	17.17
2.25	645	26.9	147	72.9	18.31
2.3	659	27.5	153.1	75.9	19.49
2.35	674	28.1	159.4	78.9	20.74
2.4	**688**	**28.7**	**165.7**	**82**	**22.02**

C	80	90	100	110	120	130	140	145	150
Fq	.80	.90	1.00	1.10	1.20	1.30	1.40	1.45	1.50
Fhc	1.511	1.215	1.000	0.838	0.713	0.615	0.536	0.503	0.472
Fhm	1.498	1.210	1.000	0.842	0.719	0.622	0.544	0.510	0.480

Multiplying factors fq, fhc and fhm for dischange, head loss and modified head loss respectively wrt C values.

TABLE: HW 08 DISCHARGE AND LOSS OF HEAD AND POWER IN PIPES
(Hazen-Williams Formula, C=100)
(Modified Hazen-Williams Formula, CR=1)

Diameter of pipe 80 mm **Area 5.0266E-03 sqm**

Velocity V, m/s	*Flow kL/d*	*Flow kL/h*	*Head Loss Hfc,m/km*	*Head Loss Hfm,m/km*	*kW Wasted/km 60% efficiency*
1	2	3	4	5	6
0.4	174	7.3	4.8	2.5	0.16
0.45	195	8.1	5.8	3.1	0.22
0.5	**217**	**9**	**7**	**3.7**	**0.29**
0.55	239	10	8.6	4.5	0.4
0.6	261	10.9	10	5.2	0.5
0.65	282	11.8	11.6	6	0.63
0.7	304	12.7	13.3	6.9	0.78
0.75	326	13.6	15.1	7.8	0.95
0.8	347	14.5	17	8.8	1.14
0.85	369	15.4	19	9.8	1.35
0.9	391	16.3	21.1	10.9	1.59
0.95	413	17.2	23.4	12	1.86
1	**434**	**18.1**	**25.7**	**13.1**	**2.15**
1.05	456	19	28.1	14.3	2.47
1.1	478	19.9	30.6	15.6	2.82
1.15	499	20.8	33.2	16.9	3.2
1.2	521	21.7	35.9	18.2	3.61
1.25	543	22.6	38.7	19.6	4.05
1.3	565	23.5	41.6	21	4.53
1.35	586	24.4	44.6	22.5	5.04
1.4	608	25.3	47.7	24	5.59
1.45	630	26.3	51.3	25.8	6.25
1.5	**651**	**27.1**	**54.2**	**27.2**	**6.8**
1.55	673	28	57.6	28.9	7.47
1.6	695	29	61.5	30.8	8.26
1.65	717	29.9	65	32.5	9
1.7	738	30.8	68.7	34.3	9.8
1.75	760	31.7	72.5	36.2	10.64
1.8	782	32.6	76.3	38	11.52
1.85	803	33.5	80.3	40	12.45
1.9	825	34.4	84.3	41.9	13.43
1.95	847	35.3	88.5	43.9	14.46
2	**869**	**36.2**	**92.7**	**46**	**15.54**
2.05	890	37.1	97	48.1	16.66
2.1	912	38	101.4	50.2	17.84
2.15	934	38.9	105.9	52.4	19.07
2.2	955	39.8	110.5	54.6	20.36
2.25	977	40.7	115.1	56.9	21.69
2.3	999	41.6	119.9	59.1	23.09
2.35	1021	42.5	124.7	61.5	24.54
2.4	**1042**	**43.4**	**129.7**	**63.9**	**26.06**

C	80	90	100	110	120	130	140	145	150
Fq	.80	.90	1.00	1.10	1.20	1.30	1.40	1.45	1.50
Fhc	1.511	1.215	1.000	0.838	0.713	0.615	0.536	0.503	0.472
Fhm	1.498	1.210	1.000	0.842	0.719	0.622	0.544	0.510	0.480

Multiplying factors fq, fhc and fhm for dischange, head loss and modified head loss respectively wrt C values.

TABLE: HW 09 DISCHARGE AND LOSS OF HEAD AND POWER IN PIPES
(Hazen-Williams Formula, C=100)
(Modified Hazen-Williams Formula, CR=1)

Diameter of pipe 100 mm **Area 7.854E-3 sqm**

Velocity V, m/s	*Flow kL/d*	*Flow kL/h*	*Head Loss Hfc,m/km*	*Head Loss Hfm,m/km*	*kW Wasted/km 60% efficiency*
1	2	3	4	5	6
0.4	271	11.3	3.6	1.9	0.19
0.45	305	12.7	4.5	2.4	0.26
0.5	**339**	**14.1**	**5.5**	**2.9**	**0.36**
0.55	373	15.5	6.5	3.4	0.47
0.6	407	17	7.7	4	0.61
0.65	441	18.4	8.9	4.6	0.76
0.7	475	19.8	10.2	5.3	0.94
0.75	509	21.2	11.6	6	1.14
0.8	543	22.6	13.1	6.7	1.37
0.85	577	24	14.6	7.5	1.62
0.9	611	25.5	16.3	8.3	1.92
0.95	645	26.9	18	9.2	2.24
1	**679**	**28.3**	**19.8**	**10.1**	**2.59**
1.05	713	29.7	21.7	11	2.98
1.1	746	31.1	23.6	11.9	3.4
1.15	780	32.5	25.6	12.9	3.85
1.2	814	33.9	27.7	14	4.35
1.25	848	35.3	29.8	15	4.87
1.3	882	36.8	32.2	16.2	5.49
1.35	916	38.2	34.5	17.3	6.1
1.4	950	39.6	36.9	18.5	6.77
1.45	984	41	39.4	19.7	7.48
1.5	**1018**	**42.4**	**41.9**	**20.9**	**8.23**
1.55	1052	43.8	44.5	22.2	9.02
1.6	1086	45.3	47.4	23.6	9.94
1.65	1120	46.7	50.1	24.9	10.83
1.7	1154	48.1	52.9	26.3	11.78
1.75	1188	49.5	55.8	27.7	12.79
1.8	1221	50.9	58.8	29.1	13.86
1.85	1255	52.3	61.8	30.6	14.96
1.9	1289	53.7	64.9	32.1	16.14
1.95	1323	55.1	68.1	33.6	17.37
2	**1357**	**56.5**	**71.3**	**35.2**	**18.65**
2.05	1391	58	74.8	36.9	20.09
2.1	1425	59.4	78.2	38.5	21.51
2.15	1459	60.8	81.7	40.2	23
2.2	1493	62.2	85.2	41.9	24.54
2.25	1527	63.6	88.8	43.6	26.15
2.3	1561	65	92.4	45.4	27.81
2.35	1595	66.5	96.4	47.3	29.68
2.4	**1629**	**67.9**	**100.2**	**49.1**	**31.5**

C	80	90	100	110	120	130	140	145	150
Fq	.80	.90	1.00	1.10	1.20	1.30	1.40	1.45	1.50
Fhc	1.511	1.215	1.000	0.838	0.713	0.615	0.536	0.503	0.472
Fhm	1.498	1.210	1.000	0.842	0.719	0.622	0.544	0.510	0.480

Multiplying factors fq, fhc and fhm for dischange, head loss and modified head loss respectively wrt C values.

TABLE: HW 10 DISCHARGE AND LOSS OF HEAD AND POWER IN PIPES
(Hazen-Williams Formula, C=100)
(Modified Hazen-Williams Formula, CR=1)

Diameter of pipe 125 mm **Area 1.2272E-02 sqm**

Velocity V, m/s	*Flow kL/d*	*Flow kL/h*	*Head Loss Hfc,m/km*	*Head Loss Hfm,m/km*	*kW Wasted/km 60% efficiency*
1	2	3	4	5	6
0.4	424	17.7	2.8	1.5	0.23
0.45	477	19.9	3.5	1.8	0.32
0.5	**530**	**22.1**	**4.2**	**2.2**	**0.43**
0.55	583	24.3	5	2.6	0.56
0.6	636	26.5	5.9	3.1	0.72
0.65	689	28.7	6.9	3.5	0.92
0.7	742	30.9	7.9	4	1.13
0.75	795	33.1	8.9	4.6	1.36
0.8	848	35.3	10.1	5.1	1.65
0.85	901	37.5	11.3	5.7	1.96
0.9	954	39.8	12.6	6.4	2.32
0.95	1007	42	13.9	7	2.7
1	**1060**	**44.2**	**15.3**	**7.7**	**3.13**
1.05	1113	46.4	16.7	8.4	3.59
1.1	1166	48.6	18.2	9.2	4.1
1.15	1219	50.8	19.8	9.9	4.66
1.2	1272	53	21.4	10.7	5.25
1.25	1325	55.2	23	11.5	5.88
1.3	1378	57.4	24.8	12.4	6.59
1.35	1431	59.6	26.6	13.3	7.34
1.4	1484	61.8	28.4	14.2	8.13
1.45	1537	64	30.3	15.1	8.98
1.5	**1590**	**66.3**	**32.3**	**16.1**	**9.92**
1.55	1643	68.5	34.4	17	10.91
1.6	1697	70.7	36.4	18.1	11.92
1.65	1750	72.9	38.6	19.1	13.03
1.7	1803	75.1	40.7	20.1	14.15
1.75	1856	77.3	43	21.2	15.39
1.8	1909	79.5	45.3	22.3	16.67
1.85	1962	81.8	47.7	23.5	18.07
1.9	2015	84	50.1	24.7	19.48
1.95	2068	86.2	52.6	25.8	20.99
2	**2121**	**88.4**	**55.1**	**27.1**	**22.55**
2.05	2174	90.6	57.7	28.3	24.2
2.1	2227	92.8	60.3	29.5	25.91
2.15	2280	95	63	30.8	27.71
2.2	2333	97.2	65.7	32.1	29.57
2.25	2386	99.4	68.5	33.4	31.53
2.3	2439	101.6	71.3	34.8	33.54
2.35	2492	103.8	74.2	36.2	35.66
2.4	**2545**	**106**	**77.1**	**37.6**	**37.84**

C	80	90	100	110	120	130	140	145	150
Fq	.80	.90	1.00	1.10	1.20	1.30	1.40	1.45	1.50
Fhc	1.511	1.215	1.000	0.838	0.713	0.615	0.536	0.503	0.472
Fhm	1.498	1.210	1.000	0.842	0.719	0.622	0.544	0.510	0.480

Multiplying factors fq, fhc and fhm for dischange, head loss and modified head loss respectively wrt C values.

TABLE: HW 11 DISCHARGE AND LOSS OF HEAD AND POWER IN PIPES
(Hazen-Williams Formula, C=100)
(Modified Hazen-Williams Formula, CR=1)

Diameter of pipe 150 mm **Area 1.7672E-2 sqm**

Velocity V, m/s	*Flow kL/d*	*Flow kL/h*	*Head Loss Hfc,m/km*	*Head Loss Hfm,m/km*	*kW Wasted/km 60% efficiency*
1	2	3	4	5	6
0.4	611	25.5	2.3	1.2	0.27
0.45	687	28.6	2.8	1.5	0.37
0.5	**763**	**31.8**	**3.4**	**1.8**	**0.5**
0.55	840	35	4.1	2.1	0.66
0.6	916	38.2	4.8	2.5	0.85
0.65	992	41.3	5.5	2.8	1.05
0.7	1069	44.5	6.4	3.2	1.32
0.75	1145	47.7	7.2	3.7	1.59
0.8	1221	50.9	8.2	4.1	1.93
0.85	1298	54.1	9.1	4.6	2.28
0.9	1374	57.3	10.2	5.1	2.71
0.95	1451	60.5	11.2	5.7	3.14
1	**1527**	**63.6**	**12.3**	**6.2**	**3.62**
1.05	1603	66.8	13.5	6.8	4.18
1.1	1680	70	14.7	7.4	4.76
1.15	1756	73.2	16	8	5.42
1.2	1832	76.3	17.3	8.6	6.11
1.25	1909	79.5	18.6	9.3	6.85
1.3	1985	82.7	20	10	7.66
1.35	2061	85.9	21.5	10.7	8.55
1.4	2138	89.1	23	11.4	9.49
1.45	2214	92.3	24.6	12.2	10.51
1.5	**2290**	**95.4**	**26.1**	**12.9**	**11.53**
1.55	2367	98.6	27.8	13.7	12.69
1.6	2443	101.8	29.5	14.5	13.9
1.65	2519	105	31.2	15.4	15.17
1.7	2596	108.2	33	16.2	16.53
1.75	2672	111.3	34.7	17.1	17.88
1.8	2748	114.5	36.6	18	19.4
1.85	2825	117.7	38.5	18.9	20.98
1.9	2901	120.9	40.5	19.8	22.67
1.95	2977	124	42.4	20.8	24.34
2	**3054**	**127.3**	**44.6**	**21.8**	**26.29**
2.05	3130	130.4	46.6	22.7	28.13
2.1	3206	133.6	48.7	23.8	30.12
2.15	3283	136.8	50.9	24.8	32.24
2.2	3359	140	53.1	25.9	34.42
2.25	3435	143.1	55.3	26.9	36.64
2.3	3512	146.3	57.6	28	39.02
2.35	3588	149.5	60	29.1	41.53
2.4	**3664**	**152.7**	**62.4**	**30.3**	**44.12**

C	80	90	100	110	120	130	140	145	150
Fq	.80	.90	1.00	1.10	1.20	1.30	1.40	1.45	1.50
Fhc	1.511	1.215	1.000	0.838	0.713	0.615	0.536	0.503	0.472
Fhm	1.498	1.210	1.000	0.842	0.719	0.622	0.544	0.510	0.480

Multiplying factors fq, fhc and fhm for dischange, head loss and modified head loss respectively wrt **C** values.

TABLE: HW 12 DISCHARGE AND LOSS OF HEAD AND POWER IN PIPES
(Hazen-Williams Formula, C=100)
(Modified Hazen-Williams Formula, CR=1)

Diameter of pipe 170 mm **Area 2.2698E-02 sqm**

Velocity V, m/s	*Flow kL/d*	*Flow kL/h*	*Head Loss Hfc,m/km*	*Head Loss Hfm,m/km*	*kW Wasted/km 60% efficiency*
1	2	3	4	5	6
0.4	784	32.7	2	1	0.3
0.45	883	36.8	2.4	1.3	0.41
0.5	**981**	**40.9**	**3**	**1.5**	**0.57**
0.55	1079	45	3.5	1.8	0.73
0.6	1177	49	4.1	2.1	0.93
0.65	1275	53.1	4.8	2.5	1.18
0.7	1373	57.2	5.5	2.8	1.46
0.75	1471	61.3	6.3	3.2	1.79
0.8	1569	65.4	7.1	3.6	2.15
0.85	1667	69.5	7.9	4	2.54
0.9	1765	73.5	8.8	4.4	2.99
0.95	1863	77.6	9.7	4.9	3.49
1	**1961**	**81.7**	**10.7**	**5.3**	**4.05**
1.05	2059	85.8	11.7	5.8	4.65
1.1	2157	89.9	12.7	6.4	5.29
1.15	2255	94	13.8	6.9	6.01
1.2	2353	98	14.9	7.4	6.76
1.25	2451	102.1	16.1	8	7.61
1.3	2550	106.3	17.3	8.6	8.51
1.35	2648	110.3	18.6	9.2	9.5
1.4	2746	114.4	19.9	9.8	10.54
1.45	2844	118.5	21.2	10.5	11.63
1.5	**2942**	**122.6**	**22.6**	**11.1**	**12.83**
1.55	3040	126.7	24	11.8	14.08
1.6	3138	130.8	25.5	12.5	15.44
1.65	3236	134.8	26.9	13.2	16.79
1.7	3334	138.9	28.5	14	18.33
1.75	3432	143	30	14.7	19.86
1.8	3530	147.1	31.7	15.5	21.59
1.85	3628	151.2	33.3	16.3	23.31
1.9	3726	155.3	35	17.1	25.17
1.95	3824	159.3	36.7	17.9	27.07
2	**3922**	**163.4**	**38.5**	**18.7**	**29.13**
2.05	4020	167.5	40.3	19.6	31.25
2.1	4118	171.6	42.1	20.5	33.45
2.15	4216	175.7	44	21.4	35.79
2.2	4315	179.8	45.9	22.3	38.21
2.25	4413	183.9	47.9	23.2	40.78
2.3	4511	188	49.9	24.2	43.43
2.35	4609	192	51.8	25.1	46.05
2.4	**4707**	**196.1**	**53.9**	**26.1**	**48.94**

C	80	90	100	110	120	130	140	145	150
Fq	.80	.90	1.00	1.10	1.20	1.30	1.40	1.45	1.50
Fhc	1.511	1.215	1.000	0.838	0.713	0.615	0.536	0.503	0.472
Fhm	1.498	1.210	1.000	0.842	0.719	0.622	0.544	0.510	0.480

Multiplying factors fq, fhc and fhm for dischange, head loss and modified head loss respectively wrt C values.

TABLE: HW 13 DISCHARGE AND LOSS OF HEAD AND POWER IN PIPES
(Hazen-Williams Formula, C=100)
(Modified Hazen-Williams Formula, CR=1)

Diameter of pipe 190 mm **Area 2.8353E-02 sqm**

Velocity V, m/s	*Flow kL/d*	*Flow kL/h*	*Head Loss Hfc,m/km*	*Head Loss Hfm,m/km*	*kW Wasted/km 60% efficiency*
1	2	*3*	4	5	6
0.4	980	40.8	1.7	0.9	0.32
0.45	1102	45.9	2.1	1.1	0.45
0.5	**1225**	**51**	**2.6**	**1.3**	**0.61**
0.55	1347	56.1	3.1	1.6	0.81
0.6	1470	61.3	3.6	1.9	1.02
0.65	1592	66.3	4.2	2.1	1.29
0.7	1715	71.5	4.8	2.5	1.59
0.75	1837	76.5	5.5	2.8	1.95
0.8	1960	81.7	6.2	3.1	2.35
0.85	2082	86.8	6.9	3.5	2.77
0.9	2205	91.9	7.7	3.9	3.28
0.95	2327	97	8.5	4.3	3.82
1	**2450**	**102.1**	**9.4**	**4.7**	**4.44**
1.05	2572	107.2	10.2	5.1	5.06
1.1	2695	112.3	11.2	5.6	5.82
1.15	2817	117.4	12.1	6	6.58
1.2	2940	122.5	13.1	6.5	7.43
1.25	3062	127.6	14.2	7	8.39
1.3	3185	132.7	15.2	7.5	9.34
1.35	3307	137.8	16.3	8.1	10.4
1.4	3430	142.9	17.5	8.6	11.58
1.45	3552	148	18.6	9.2	12.75
1.5	**3675**	**153.1**	**19.8**	**9.8**	**14.04**
1.55	3797	158.2	21.1	10.4	15.46
1.6	3920	163.3	22.3	11	16.86
1.65	4042	168.4	23.7	11.6	18.48
1.7	4165	173.5	25	12.2	20.08
1.75	4287	178.6	26.4	12.9	21.83
1.8	4410	183.8	27.8	13.6	23.66
1.85	4532	188.8	29.2	14.3	25.53
1.9	4655	194	30.7	15	27.58
1.95	4777	199	32.2	15.7	29.67
2	**4899**	**204.1**	**33.8**	**16.4**	**31.94**
2.05	5022	209.3	35.4	17.2	34.3
2.1	5144	214.3	37	17.9	36.71
2.15	5267	219.5	38.6	18.7	39.23
2.2	5389	224.5	40.3	19.5	41.89
2.25	5512	229.7	42	20.3	44.67
2.3	5634	234.8	43.8	21.2	47.62
2.35	5757	239.9	45.6	22	50.65
2.4	**5879**	**245**	**47.4**	**22.8**	**53.77**

C	80	90	100	110	120	130	140	145	150
Fq	.80	.90	1.00	1.10	1.20	1.30	1.40	1.45	1.50
Fhc	1.511	1.215	1.000	0.838	0.713	0.615	0.536	0.503	0.472
Fhm	1.498	1.210	1.000	0.842	0.719	0.622	0.544	0.510	0.480

Multiplying factors fq, fhc and fhm for dischange, head loss and modified head loss respectively wrt C values.

TABLE: HW 14 DISCHARGE AND LOSS OF HEAD AND POWER IN PIPES
(Hazen-Williams Formula, C=100)
(Modified Hazen-Williams Formula, CR=1)

Diameter of pipe 200 mm **Area 3.1416E-02 sqm**

Velocity V, m/s	*Flow kL/d*	*Flow kL/h*	*Head Loss Hfc,m/km*	*Head Loss Hfm,m/km*	*kW Wasted/km 60% efficiency*
1	*2*	*3*	*4*	*5*	*6*
0.4	1086	45.3	1.6	0.8	0.34
0.45	1221	50.9	2	1	0.47
0.5	**1357**	**56.5**	**2.4**	**1.3**	**0.63**
0.55	1493	62.2	2.9	1.5	0.84
0.6	1629	67.9	3.4	1.7	1.07
0.65	1764	73.5	4	2	1.36
0.7	1900	79.2	4.6	2.3	1.69
0.75	2036	84.8	5.2	2.6	2.04
0.8	2172	90.5	5.8	2.9	2.43
0.85	2307	96.1	6.5	3.3	2.89
0.9	2443	101.8	7.3	3.6	3.44
0.95	2579	107.5	8	4	3.98
1	**2714**	**113.1**	**8.8**	**4.4**	**4.61**
1.05	2850	118.8	9.7	4.8	5.34
1.1	2986	124.4	10.5	5.2	6.05
1.15	3122	130.1	11.4	5.7	6.87
1.2	3257	135.7	12.4	6.1	7.79
1.25	3393	141.4	13.3	6.6	8.71
1.3	3529	147	14.3	7.1	9.73
1.35	3664	152.7	15.4	7.6	10.89
1.4	3800	158.3	16.4	8.1	12.02
1.45	3936	164	17.5	8.6	13.29
1.5	**4072**	**169.7**	**18.7**	**9.2**	**14.69**
1.55	4207	175.3	19.9	9.7	16.15
1.6	4343	181	21.1	10.3	17.68
1.65	4479	186.6	22.3	10.9	19.27
1.7	4614	192.3	23.6	11.5	21.01
1.75	4750	197.9	24.8	12.1	22.72
1.8	4886	203.6	26.2	12.8	24.7
1.85	5022	209.3	27.6	13.4	26.75
1.9	5157	214.9	28.9	14.1	28.76
1.95	5293	220.5	30.4	14.8	31.04
2	**5429**	**226.2**	**31.8**	**15.4**	**33.3**
2.05	5565	231.9	33.3	16.2	35.75
2.1	5700	237.5	34.8	16.9	38.27
2.15	5836	243.2	36.4	17.6	40.99
2.2	5972	248.8	38	18.4	43.77
2.25	6107	254.5	39.6	19.1	46.66
2.3	6243	260.1	41.2	19.9	49.62
2.35	6379	265.8	42.9	20.7	52.8
2.4	**6515**	**271.5**	**44.6**	**21.5**	**56.06**

C	80	90	100	110	120	130	140	145	150
Fq	.80	.90	1.00	1.10	1.20	1.30	1.40	1.45	1.50
Fhc	1.511	1.215	1.000	0.838	0.713	0.615	0.536	0.503	0.472
Fhm	1.498	1.210	1.000	0.842	0.719	0.622	0.544	0.510	0.480

Multiplying factors fq, fhc and fhm for dischange, head loss and modified head loss respectively wrt C values.

TABLE: HW 15 DISCHARGE AND LOSS OF HEAD AND POWER IN PIPES
(Hazen-Williams Formula, C=100)
(Modified Hazen-Williams Formula, CR=1)

Diameter of pipe 250 mm **Area 4.9088E-02 sqm**

Velocity V, m/s	*Flow ML/d*	*Flow kL/min*	*Head Loss Hfc,m/km*	*Head Loss Hfm,m/km*	*kW Wasted/km 60% efficiency*
1	2	3	4	5	6
0.4	1.7	1.2	1.3	0.7	0.43
0.45	1.9	1.3	1.5	0.8	0.54
0.5	**2.1**	**1.5**	**1.9**	**1**	**0.79**
0.55	2.3	1.6	2.2	1.1	0.98
0.6	2.5	1.7	2.5	1.2	1.18
0.65	2.8	1.9	3	1.5	1.58
0.7	3	2.1	3.6	1.8	2.1
0.75	3.2	2.2	4	2	2.44
0.8	3.4	2.4	4.7	2.3	3.13
0.85	3.6	2.5	5	2.5	3.47
0.9	3.8	2.6	5.4	2.7	3.9
0.95	4	2.8	6.2	3.1	4.82
1	**4.2**	**2.9**	**6.6**	**3.3**	**5.32**
1.05	4.5	3.1	7.5	3.7	6.46
1.1	4.7	3.3	8.4	4.2	7.7
1.15	4.9	3.4	8.9	4.4	8.41
1.2	5.1	3.5	9.4	4.6	9.14
1.25	5.3	3.7	10.4	5.1	10.69
1.3	5.5	3.8	10.9	5.4	11.51
1.35	5.7	4	12	5.9	13.33
1.4	5.9	4.1	12.5	6.1	14.24
1.45	6.1	4.2	13.1	6.4	15.28
1.5	**6.4**	**4.4**	**14.3**	**7**	**17.48**
1.55	6.6	4.6	15.5	7.6	19.81
1.6	6.8	4.7	16.2	7.9	21.15
1.65	7	4.9	17.4	8.5	23.68
1.7	7.2	5	18.1	8.8	25.14
1.75	7.4	5.1	18.8	9.1	26.63
1.8	7.6	5.3	20.2	9.8	29.74
1.85	7.8	5.4	20.9	10.1	31.35
1.9	8.1	5.6	22.3	10.8	34.69
1.95	8.3	5.8	23.8	11.5	38.34
2	**8.5**	**5.9**	**24.6**	**11.9**	**40.32**
2.05	8.7	6	25.4	12.2	42.33
2.1	8.9	6.2	27	13	46.5
2.15	9.1	6.3	27.8	13.4	48.65
2.2	9.3	6.5	29.4	14.2	53.08
2.25	9.5	6.6	30.3	14.6	55.55
2.3	9.8	6.8	32	15.4	60.44
2.35	10	6.9	32.9	15.8	63.06
2.4	**10.2**	**7.1**	**34.7**	**16.6**	**68.44**

C	80	90	100	110	120	130	140	145	150
Fq	.80	.90	1.00	1.10	1.20	1.30	1.40	1.45	1.50
Fhc	1.511	1.215	1.000	0.838	0.713	0.615	0.536	0.503	0.472
Fhm	1.498	1.210	1.000	0.842	0.719	0.622	0.544	0.510	0.480

Multiplying factors fq, fhc and fhm for dischange, head loss and modified head loss respectively wrt C values.

TABLE: HW 16 DISCHARGE AND LOSS OF HEAD AND POWER IN PIPES
(Hazen-Williams Formula, C=100)
(Modified Hazen-Williams Formula, CR=1)

Diameter of pipe 270 mm **Area 5.7256E-02 sqm**

Velocity V, m/s	*Flow ML/d*	*Flow kL/min*	*Head Loss Hfc,m/km*	*Head Loss Hfm,m/km*	*kW Wasted/km 60% efficiency*
1	2	3	4	5	6
0.4	2	1.4	1.2	0.6	0.47
0.45	2.2	1.5	1.3	0.7	0.54
0.5	**2.5**	**1.7**	**1.7**	**0.9**	**0.8**
0.55	2.7	1.9	2.1	1.1	1.11
0.6	3	2.1	2.5	1.3	1.46
0.65	3.2	2.2	2.7	1.4	1.65
0.7	3.5	2.4	3.2	1.6	2.13
0.75	3.7	2.6	3.7	1.9	2.67
0.8	4	2.8	4.3	2.1	3.34
0.85	4.2	2.9	4.5	2.3	3.63
0.9	4.5	3.1	5.1	2.6	4.39
0.95	4.7	3.3	5.8	2.9	5.32
1	**4.9**	**3.4**	**6.1**	**3**	**5.76**
1.05	5.2	3.6	6.8	3.4	6.8
1.1	5.4	3.8	7.5	3.7	7.92
1.15	5.7	4	8.2	4.1	9.11
1.2	5.9	4.1	8.6	4.2	9.79
1.25	6.2	4.3	9.4	4.6	11.23
1.3	6.4	4.4	9.8	4.8	11.98
1.35	6.7	4.7	11.1	5.4	14.49
1.4	6.9	4.8	11.5	5.6	15.33
1.45	7.2	5	12.5	6.1	17.36
1.5	**7.4**	**5.1**	**12.9**	**6.3**	**18.28**
1.55	7.7	5.3	13.9	6.8	20.46
1.6	7.9	5.5	14.9	7.2	22.76
1.65	8.2	5.7	15.9	7.7	25.18
1.7	8.4	5.8	16.4	8	26.42
1.75	8.7	6	17.5	8.5	29.17
1.8	8.9	6.2	18.5	9	31.86
1.85	9.2	6.4	19.7	9.5	35.02
1.9	9.4	6.5	20.2	9.8	36.47
1.95	9.6	6.7	21.4	10.3	39.83
2	**9.9**	**6.9**	**22.6**	**10.9**	**43.32**
2.05	10.1	7	23.2	11.2	45.11
2.1	10.4	7.2	24.5	11.8	49
2.15	10.6	7.4	25.7	12.4	52.83
2.2	10.9	7.6	27	13	57
2.25	11.1	7.7	27.7	13.3	59.25
2.3	11.4	7.9	29	13.9	63.64
2.35	11.6	8.1	30.4	14.6	68.4
2.4	**11.9**	**8.3**	**31.8**	**15.2**	**73.32**

C	80	90	100	110	120	130	140	145	150
Fq	.80	.90	1.00	1.10	1.20	1.30	1.40	1.45	1.50
Fhc	1.511	1.215	1.000	0.838	0.713	0.615	0.536	0.503	0.472
Fhm	1.498	1.210	1.000	0.842	0.719	0.622	0.544	0.510	0.480

Multiplying factors fq, fhc and fhm for dischange, head loss and modified head loss respectively wrt C values.

TABLE: HW 17 DISCHARGE AND LOSS OF HEAD AND POWER IN PIPES
(Hazen-Williams Formula, C=100)
(Modified Hazen-Williams Formula, CR=1)

Diameter of pipe 300 mm **Area 7.0686E-02 sqm**

Velocity V, m/s	*Flow ML/d*	*Flow kL/min*	*Head Loss Hfc,m/km*	*Head Loss Hfm,m/km*	*kW Wasted/km 60% efficiency*
1	*2*	*3*	*4*	*5*	*6*
0.4	2.4	1.7	1	0.5	0.47
0.45	2.7	1.9	1.2	0.6	0.63
0.5	**3.1**	**2.2**	**1.6**	**0.8**	**0.98**
0.55	3.4	2.4	1.9	1	1.27
0.6	3.7	2.6	2.2	1.1	1.59
0.65	4	2.8	2.5	1.3	1.94
0.7	4.3	3	2.9	1.5	2.42
0.75	4.6	3.2	3.3	1.6	2.93
0.8	4.9	3.4	3.6	1.8	3.4
0.85	5.2	3.6	4.1	2	4.1
0.9	5.5	3.8	4.5	2.2	4.75
0.95	5.8	4	4.9	2.4	5.44
1	**6.1**	**4.2**	**5.4**	**2.7**	**6.3**
1.05	6.4	4.4	5.9	2.9	7.21
1.1	6.7	4.7	6.6	3.3	8.62
1.15	7	4.9	7.2	3.5	9.8
1.2	7.3	5.1	7.7	3.8	10.91
1.25	7.6	5.3	8.3	4.1	12.22
1.3	7.9	5.5	8.9	4.4	13.6
1.35	8.2	5.7	9.5	4.6	15.04
1.4	8.6	6	10.4	5.1	17.33
1.45	8.9	6.2	11.1	5.4	19.12
1.5	**9.2**	**6.4**	**11.8**	**5.7**	**20.98**
1.55	9.5	6.6	12.5	6.1	22.92
1.6	9.8	6.8	13.2	6.4	24.93
1.65	10.1	7	13.9	6.7	27.03
1.7	10.4	7.2	14.6	7.1	29.2
1.75	10.7	7.4	15.4	7.4	31.66
1.8	11	7.6	16.2	7.8	34.2
1.85	11.3	7.8	17	8.2	36.83
1.9	11.6	8.1	18.2	8.8	40.95
1.95	11.9	8.3	19.1	9.2	44.04
2	**12.2**	**8.5**	**19.9**	**9.6**	**46.99**
2.05	12.5	8.7	20.8	10	50.27
2.1	12.8	8.9	21.7	10.4	53.65
2.15	13.1	9.1	22.6	10.8	57.13
2.2	13.4	9.3	23.5	11.3	60.71
2.25	13.7	9.5	24.5	11.7	64.65
2.3	14	9.7	25.4	12.2	68.44
2.35	14.4	10	26.9	12.8	74.72
2.4	**14.7**	**10.2**	**27.9**	**13.3**	**79.05**

C	80	90	100	110	120	130	140	145	150
Fq	.80	.90	1.00	1.10	1.20	1.30	1.40	1.45	1.50
Fhc	1.511	1.215	1.000	0.838	0.713	0.615	0.536	0.503	0.472
Fhm	1.498	1.210	1.000	0.842	0.719	0.622	0.544	0.510	0.480

Multiplying factors fq, fhc and fhm for dischange, head loss and modified head loss respectively wrt C values.

TABLE: HW 18 DISCHARGE AND LOSS OF HEAD AND POWER IN PIPES
(Hazen-Williams Formula, C=100)
(Modified Hazen-Williams Formula, CR=1)

Diameter of pipe 350 mm **Area 9.6211E-02 sqm**

Velocity V, m/s	*Flow ML/d*	*Flow kL/min*	*Head Loss Hfc,m/km*	*Head Loss Hfm,m/km*	*kW Wasted/km 60% efficiency*
1	2	3	4	5	6
0.4	3.3	2.3	0.8	0.43	0.51
0.45	3.7	2.6	1	0.53	0.72
0.5	**4.2**	**2.9**	**1.3**	**0.65**	**1.05**
0.55	4.6	3.2	1.5	0.78	1.33
0.6	5	3.5	1.8	0.92	1.75
0.65	5.4	3.8	2.1	1.06	2.22
0.7	5.8	4	2.3	1.17	2.56
0.75	6.2	4.3	2.7	1.33	3.23
0.8	6.7	4.7	3.1	1.56	4.05
0.85	7.1	4.9	3.4	1.68	4.63
0.9	7.5	5.2	3.8	1.87	5.49
0.95	7.9	5.5	4.2	2.07	6.42
1	**8.3**	**5.8**	**4.6**	**2.28**	**7.41**
1.05	8.7	6	4.9	2.43	8.17
1.1	9.1	6.3	5.4	2.65	9.45
1.15	9.6	6.7	6	2.96	11.17
1.2	10	6.9	6.4	3.13	12.27
1.25	10.4	7.2	6.9	3.38	13.8
1.3	10.8	7.5	7.5	3.64	15.63
1.35	11.2	7.8	8	3.9	17.33
1.4	11.6	8.1	8.6	4.18	19.35
1.45	12.1	8.4	9.2	4.46	21.47
1.5	**12.5**	**8.7**	**9.8**	**4.76**	**23.68**
1.55	12.9	9	10.4	5.06	26
1.6	13.3	9.2	10.9	5.26	27.86
1.65	13.7	9.5	11.5	5.58	30.35
1.7	14.1	9.8	12.2	5.9	33.21
1.75	14.5	10.1	12.9	6.23	36.19
1.8	15	10.4	13.7	6.57	39.58
1.85	15.4	10.7	14.4	6.92	42.8
1.9	15.8	11	15.2	7.27	46.44
1.95	16.2	11.3	15.9	7.63	49.91
2	**16.6**	**11.5**	**16.5**	**7.88**	**52.71**
2.05	17	11.8	17.3	8.26	56.71
2.1	17.5	12.2	18.4	8.77	62.36
2.15	17.9	12.4	18.9	9.03	65.1
2.2	18.3	12.7	19.8	9.43	69.85
2.25	18.7	13	20.6	9.84	74.39
2.3	19.1	13.3	21.5	10.25	79.43
2.35	19.5	13.5	22.1	10.54	82.88
2.4	**20**	**13.9**	**23.4**	**11.11**	**90.35**

C	80	90	100	110	120	130	140	145	150
Fq	.80	.90	1.00	1.10	1.20	1.30	1.40	1.45	1.50
Fhc	1.511	1.215	1.000	0.838	0.713	0.615	0.536	0.503	0.472
Fhm	1.498	1.210	1.000	0.842	0.719	0.622	0.544	0.510	0.480

Multiplying factors fq, fhc and fhm for dischange, head loss and modified head loss respectively wrt C values.

TABLE: HW 19 DISCHARGE AND LOSS OF HEAD AND POWER IN PIPES
(Hazen-Williams Formula, C=100)
(Modified Hazen-Williams Formula, CR=1)

Diameter of pipe 400 mm **Area 0.1257 sqm**

Velocity V, m/s	*Flow ML/d*	*Flow kL/min*	*Head Loss Hfc,m/km*	*Head Loss Hfm,m/km*	*kW Wasted/km 60% efficiency*
1	*2*	*3*	*4*	*5*	*6*
0.4	4.3	3	0.7	0.36	0.58
0.45	4.9	3.4	0.9	0.46	0.85
0.5	**5.4**	**3.8**	**1.1**	**0.56**	**1.16**
0.55	6	4.2	1.3	0.67	1.52
0.6	6.5	4.5	1.5	0.76	1.88
0.65	7.1	4.9	1.8	0.89	2.45
0.7	7.6	5.3	2	1.02	2.94
0.75	8.1	5.6	2.3	1.13	3.58
0.8	8.7	6	2.6	1.28	4.33
0.85	9.2	6.4	2.9	1.44	5.16
0.9	9.8	6.8	3.2	1.6	6.04
0.95	10.3	7.2	3.6	1.78	7.2
1	**10.9**	**7.6**	**4**	**1.96**	**8.44**
1.05	11.4	7.9	4.3	2.1	9.44
1.1	11.9	8.3	4.7	2.3	10.84
1.15	12.5	8.7	5.1	2.5	12.33
1.2	13	9	5.5	2.66	13.75
1.25	13.6	9.4	5.9	2.88	15.41
1.3	14.1	9.8	6.4	3.1	17.42
1.35	14.7	10.2	6.9	3.34	19.55
1.4	15.2	10.6	7.4	3.58	21.79
1.45	15.7	10.9	7.8	3.76	23.62
1.5	**16.3**	**11.3**	**8.3**	**4.02**	**26.05**
1.55	16.8	11.7	8.9	4.28	28.93
1.6	17.4	12.1	9.4	4.55	31.59
1.65	17.9	12.4	9.9	4.75	34.1
1.7	18.5	12.8	10.5	5.03	37.33
1.75	19	13.2	11.1	5.32	40.7
1.8	19.5	13.5	11.6	5.54	43.5
1.85	20.1	14	12.4	5.92	48.22
1.9	20.6	14.3	12.9	6.15	51.24
1.95	21.2	14.7	13.5	6.47	55.13
2	**21.7**	**15.1**	**14.2**	**6.79**	**59.56**
2.05	22.3	15.5	14.9	7.12	64.15
2.1	22.8	15.8	15.5	7.37	68.03
2.15	23.3	16.2	16.2	7.71	72.9
2.2	23.9	16.6	16.9	8.06	77.93
2.25	24.4	16.9	17.5	8.32	82.15
2.3	25	17.4	18.5	8.77	89.42
2.35	25.5	17.7	19.1	9.05	93.91
2.4	**26.1**	**18.1**	**19.9**	**9.42**	**100.05**

C	80	90	100	110	120	130	140	145	150
Fq	.80	.90	1.00	1.10	1.20	1.30	1.40	1.45	1.50
Fhc	1.511	1.215	1.000	0.838	0.713	0.615	0.536	0.503	0.472
Fhm	1.498	1.210	1.000	0.842	0.719	0.622	0.544	0.510	0.480

Multiplying factors fq, fhc and fhm for dischange, head loss and modified head loss respectively wrt C values.

TABLE: HW 20 DISCHARGE AND LOSS OF HEAD AND POWER IN PIPES
(Hazen-Williams Formula, C=100)
(Modified Hazen-Williams Formula, CR=1)

Diameter of pipe 450 mm **Area 0.1590 sqm**

Velocity V, m/s	*Flow ML/d*	*Flow kL/min*	*Head Loss Hfc,m/km*	*Head Loss Hfm,m/km*	*kW Wasted/km 60% efficiency*
1	*2*	*3*	*4*	*5*	*6*
0.4	5.5	3.8	0.6	0.32	0.63
0.45	6.2	4.3	0.8	0.4	0.96
0.5	**6.9**	**4.8**	**1**	**0.48**	**1.33**
0.55	7.6	5.3	1.2	0.58	1.77
0.6	8.2	5.7	1.3	0.66	2.06
0.65	8.9	6.2	1.5	0.77	2.58
0.7	9.6	6.7	1.8	0.89	3.35
0.75	10.3	7.2	2	1.01	4
0.8	11	7.6	2.2	1.11	4.64
0.85	11.7	8.1	2.5	1.25	5.63
0.9	12.4	8.6	2.8	1.39	6.69
0.95	13.1	9.1	3.1	1.54	7.84
1	**13.7**	**9.5**	**3.4**	**1.67**	**8.97**
1.05	14.4	10	3.7	1.83	10.28
1.1	15.1	10.5	4.1	2	11.96
1.15	15.8	11	4.5	2.17	13.75
1.2	16.5	11.5	4.8	2.35	15.33
1.25	17.2	11.9	5.2	2.5	17.19
1.3	17.9	12.4	5.6	2.7	19.29
1.35	18.6	12.9	6	2.9	21.5
1.4	19.2	13.3	6.3	3.06	23.28
1.45	19.9	13.8	6.8	3.27	26.07
1.5	**20.6**	**14.3**	**7.2**	**3.49**	**28.6**
1.55	21.3	14.8	7.7	3.71	31.66
1.6	22	15.3	8.2	3.94	34.85
1.65	22.7	15.8	8.7	4.18	38.18
1.7	23.4	16.3	9.2	4.42	41.66
1.75	24	16.7	9.7	4.62	45
1.8	24.7	17.2	10.2	4.88	48.73
1.85	25.4	17.6	10.6	5.08	51.82
1.9	26.1	18.1	11.2	5.35	56.31
1.95	26.8	18.6	11.8	5.62	60.97
2	**27.5**	**19.1**	**12.4**	**5.89**	**65.79**
2.05	28.2	19.6	13	6.18	70.78
2.1	28.9	20.1	13.6	6.46	75.93
2.15	29.5	20.5	14.1	6.7	80.29
2.2	30.2	21	14.8	7	86.33
2.25	30.9	21.5	15.4	7.3	91.97
2.3	31.6	21.9	16	7.55	97.33
2.35	32.3	22.4	16.6	7.87	103.29
2.4	**33**	**22.9**	**17.3**	**8.19**	**110.05**

C	80	90	100	110	120	130	140	145	150
Fq	.80	.90	1.00	1.10	1.20	1.30	1.40	1.45	1.50
Fhc	1.511	1.215	1.000	0.838	0.713	0.615	0.536	0.503	0.472
Fhm	1.498	1.210	1.000	0.842	0.719	0.622	0.544	0.510	0.480

Multiplying factors fq, fhc and fhm for dischange, head loss and modified head loss respectively wrt C values.

TABLE: HW 21 DISCHARGE AND LOSS OF HEAD AND POWER IN PIPES (Hazen-Williams Formula, C=100) (Modified Hazen-Williams Formula, CR=1)

Diameter of pipe 500 mm **Area 0.1964 sqm**

Velocity V, m/s	*Flow ML/d*	*Flow kL/min*	*Head Loss Hfc,m/km*	*Head Loss Hfm,m/km*	*kW Wasted/km 60% efficiency*
1	*2*	*3*	*4*	*5*	*6*
0.4	6.8	4.7	0.55	0.28	0.72
0.45	7.6	5.3	0.69	0.35	1.02
0.5	**8.5**	**5.9**	**0.84**	**0.42**	**1.38**
0.55	9.3	6.5	1.01	0.5	1.82
0.6	10.2	7.1	1.19	0.59	2.35
0.65	11	7.6	1.35	0.67	2.85
0.7	11.9	8.3	1.58	0.79	3.64
0.75	12.7	8.8	1.76	0.87	4.3
0.8	13.6	9.4	1.99	0.98	5.2
0.85	14.4	10	2.24	1.1	6.22
0.9	15.3	10.6	2.49	1.22	7.33
0.95	16.1	11.2	2.76	1.35	8.59
1	**17**	**11.8**	**3.04**	**1.49**	**9.96**
1.05	17.8	12.4	3.33	1.62	11.47
1.1	18.7	13	3.63	1.77	13.11
1.15	19.5	13.5	3.9	1.89	14.63
1.2	20.4	14.2	4.28	2.08	16.88
1.25	21.2	14.7	4.56	2.21	18.62
1.3	22.1	15.3	4.91	2.38	20.87
1.35	22.9	15.9	5.28	2.55	23.32
1.4	23.8	16.5	5.65	2.72	25.9
1.45	24.6	17.1	6.04	2.91	28.69
1.5	**25.4**	**17.6**	**6.37**	**3.06**	**31.14**
1.55	26.3	18.3	6.85	3.29	34.82
1.6	27.1	18.8	7.2	3.45	37.6
1.65	28	19.4	7.63	3.65	41.12
1.7	28.8	20	8.07	3.86	44.83
1.75	29.7	20.6	8.53	4.07	48.81
1.8	30.5	21.2	8.99	4.29	52.94
1.85	31.4	21.8	9.47	4.51	57.35
1.9	32.2	22.4	9.96	4.74	61.97
1.95	33.1	23	10.46	4.97	66.83
2	**33.9**	**23.5**	**10.88**	**5.17**	**71.02**
2.05	34.8	24.2	11.49	5.45	77.24
2.1	35.6	24.7	11.93	5.66	81.85
2.15	36.5	25.3	12.48	5.91	87.71
2.2	37.3	25.9	13.03	6.16	93.74
2.25	38.2	26.5	13.59	6.42	100.04
2.3	39	27.1	14.17	6.69	106.67
2.35	39.9	27.7	14.75	6.96	113.49
2.4	**40.7**	**28.3**	**15.35**	**7.23**	**120.67**

C	80	90	100	110	120	130	140	145	150
Fq	.80	.90	1.00	1.10	1.20	1.30	1.40	1.45	1.50
Fhc	1.511	1.215	1.000	0.838	0.713	0.615	0.536	0.503	0.472
Fhm	1.498	1.210	1.000	0.842	0.719	0.622	0.544	0.510	0.480

Multiplying factors fq, fhc and fhm for dischange, head loss and modified head loss respectively wrt C values.

TABLE: HW 22 DISCHARGE AND LOSS OF HEAD AND POWER IN PIPES

(Hazen-Williams Formula, C=100)

(Modified Hazen-Williams Formula, CR=1)

Diameter of pipe 550 mm **Area 0.2376 sqm**

Velocity V, m/s	*Flow ML/d*	*Flow kL/min*	*Head Loss Hfc,m/km*	*Head Loss Hfm,m/km*	*kW Wasted/km 60% efficiency*
1	2	3	4	5	6
0.4	8.2	5.7	0.5	0.25	0.79
0.45	9.2	6.4	0.62	0.31	1.1
0.5	**10.3**	**7.2**	**0.76**	**0.38**	**1.52**
0.55	11.3	7.8	0.89	0.44	1.93
0.6	12.3	8.5	1.04	0.52	2.46
0.65	13.3	9.2	1.2	0.6	3.07
0.7	14.4	10	1.41	0.7	3.92
0.75	15.4	10.7	1.59	0.79	4.73
0.8	16.4	11.4	1.79	0.88	5.67
0.85	17.4	12.1	2	0.98	6.72
0.9	18.5	12.8	2.22	1.09	7.89
0.95	19.5	13.5	2.45	1.2	9.19
1	**20.5**	**14.2**	**2.69**	**1.31**	**10.61**
1.05	21.6	15	2.98	1.45	12.42
1.1	22.6	15.7	3.24	1.57	14.13
1.15	23.6	16.4	3.51	1.7	15.99
1.2	24.6	17.1	3.8	1.84	18.05
1.25	25.7	17.8	4.09	1.98	20.22
1.3	26.7	18.5	4.39	2.12	22.56
1.35	27.7	19.2	4.7	2.27	25.07
1.4	28.7	19.9	5.03	2.42	27.8
1.45	29.8	20.7	5.41	2.6	31.11
1.5	**30.8**	**21.4**	**5.75**	**2.76**	**34.18**
1.55	31.8	22.1	6.11	2.92	37.51
1.6	32.8	22.8	6.47	3.09	40.98
1.65	33.9	23.5	6.84	3.27	44.65
1.7	34.9	24.2	7.22	3.45	48.53
1.75	35.9	24.9	7.61	3.63	52.64
1.8	36.9	25.6	8.02	3.81	57.03
1.85	38	26.4	8.49	4.03	62.26
1.9	39	27.1	8.91	4.23	67.07
1.95	40	27.8	9.34	4.43	72.13
2	**41.1**	**28.5**	**9.78**	**4.63**	**77.43**
2.05	42.1	29.2	10.23	4.84	82.98
2.1	43.1	29.9	10.69	5.05	88.79
2.15	44.1	30.6	11.15	5.27	94.78
2.2	45.2	31.4	11.7	5.52	102.05
2.25	46.2	32.1	12.19	5.75	108.69
2.3	47.2	32.8	12.68	5.97	115.53
2.35	48.2	33.5	13.19	6.21	122.74
2.4	**49.3**	**34.2**	**13.71**	**6.44**	**130.25**

C	80	90	100	110	120	130	140	145	150
Fq	.80	.90	1.00	1.10	1.20	1.30	1.40	1.45	1.50
Fhc	1.511	1.215	1.000	0.838	0.713	0.615	0.536	0.503	0.472
Fhm	1.498	1.210	1.000	0.842	0.719	0.622	0.544	0.510	0.480

Multiplying factors fq, fhc and fhm for dischange, head loss and modified head loss respectively wrt C values.

TABLE: HW 23 DISCHARGE AND LOSS OF HEAD AND POWER IN PIPES
(Hazen-Williams Formula, C=100)
(Modified Hazen-Williams Formula, CR=1)

Diameter of pipe 600 mm **Area 0.2827 sqm**

Velocity V, m/s	*Flow ML/d*	*Flow kL/min*	*Head Loss Hfc,m/km*	*Head Loss Hfm,m/km*	*kW Wasted/km 60% efficiency*
1	*2*	*3*	*4*	*5*	*6*
0.4	9.8	6.81	0.45	0.23	0.85
0.45	11	7.64	0.56	0.28	1.19
0.5	**12.2**	**8.47**	**0.68**	**0.34**	**1.6**
0.55	13.4	9.31	0.81	0.4	2.09
0.6	14.7	10.21	0.96	0.48	2.72
0.65	15.9	11.04	1.11	0.55	3.4
0.7	17.1	11.88	1.27	0.63	4.19
0.75	18.3	12.71	1.43	0.71	5.05
0.8	19.5	13.54	1.61	0.79	6.06
0.85	20.8	14.44	1.82	0.89	7.3
0.9	22	15.28	2.02	0.99	8.57
0.95	23.2	16.11	2.23	1.09	9.98
1	**24.4**	**16.94**	**2.44**	**1.19**	**11.48**
1.05	25.7	17.85	2.69	1.31	13.34
1.1	26.9	18.68	2.93	1.42	15.2
1.15	28.1	19.51	3.17	1.54	17.18
1.2	29.3	20.35	3.43	1.66	19.39
1.25	30.5	21.18	3.69	1.78	21.71
1.3	31.8	22.08	3.99	1.92	24.47
1.35	33	22.92	4.28	2.05	27.25
1.4	34.2	23.75	4.57	2.19	30.15
1.45	35.4	24.58	4.87	2.33	33.25
1.5	**36.6**	**25.42**	**5.18**	**2.48**	**36.58**
1.55	37.9	26.32	5.52	2.64	40.36
1.6	39.1	27.15	5.85	2.79	44.12
1.65	40.3	27.99	6.19	2.95	48.13
1.7	41.5	28.82	6.53	3.11	52.28
1.75	42.8	29.72	6.92	3.29	57.13
1.8	44	30.56	7.28	3.46	61.8
1.85	45.2	31.39	7.65	3.63	66.7
1.9	46.4	32.22	8.03	3.81	71.87
1.95	47.6	33.06	8.43	3.99	77.42
2	**48.9**	**33.96**	**8.86**	**4.19**	**83.58**
2.05	50.1	34.79	9.26	4.37	89.49
2.1	51.3	35.63	9.68	4.57	95.81
2.15	52.5	36.46	10.1	4.76	102.29
2.2	53.7	37.29	10.53	4.96	109.07
2.25	55	38.19	11.01	5.18	116.8
2.3	56.2	39.03	11.46	5.39	124.25
2.35	57.4	39.86	11.91	5.59	131.87
2.4	**58.6**	**40.69**	**12.38**	**5.81**	**139.93**

C	80	90	100	110	120	130	140	145	150
Fq	.80	.90	1.00	1.10	1.20	1.30	1.40	1.45	1.50
Fhc	1.511	1.215	1.000	0.838	0.713	0.615	0.536	0.503	0.472
Fhm	1.498	1.210	1.000	0.842	0.719	0.622	0.544	0.510	0.480

Multiplying factors fq, fhc and fhm for dischange, head loss and modified head loss respectively wrt C values.

TABLE: HW 24 DISCHARGE AND LOSS OF HEAD AND POWER IN PIPES
(Hazen-Williams Formula, C=100)
(Modified Hazen-Williams Formula, CR=1)

Diameter of pipe 650 mm **Area 0.3318 sqm**

Velocity V, m/s	*Flow ML/d*	*Flow kL/min*	*Head Loss Hfc,m/km*	*Head Loss Hfm,m/km*	*kW Wasted/km 60% efficiency*
1	*2*	*3*	*4*	*5*	*6*
0.4	11.5	7.99	0.41	0.21	0.91
0.45	12.9	8.96	0.51	0.26	1.27
0.5	**14.3**	**9.93**	**0.62**	**0.31**	**1.71**
0.55	15.8	10.97	0.74	0.37	2.25
0.6	17.2	11.94	0.87	0.43	2.89
0.65	18.6	12.92	1	0.5	3.59
0.7	20.1	13.96	1.16	0.57	4.5
0.75	21.5	14.93	1.31	0.64	5.43
0.8	22.9	15.9	1.47	0.72	6.49
0.85	24.4	16.94	1.65	0.81	7.76
0.9	25.8	17.92	1.84	0.9	9.16
0.95	27.2	18.89	2.02	0.99	10.6
1	**28.7**	**19.93**	**2.23**	**1.09**	**12.35**
1.05	30.1	20.9	2.44	1.18	14.17
1.1	31.5	21.88	2.66	1.29	16.17
1.15	33	22.92	2.9	1.4	18.46
1.2	34.4	23.89	3.13	1.51	20.77
1.25	35.8	24.86	3.37	1.62	23.27
1.3	37.3	25.9	3.63	1.74	26.12
1.35	38.7	26.88	3.89	1.87	29.05
1.4	40.1	27.85	4.15	1.99	32.1
1.45	41.6	28.89	4.44	2.13	35.63
1.5	**43**	**29.86**	**4.73**	**2.26**	**39.23**
1.55	44.4	30.83	5.01	2.39	42.91
1.6	45.9	31.88	5.33	2.54	47.2
1.65	47.3	32.85	5.64	2.68	51.47
1.7	48.7	33.82	5.95	2.83	55.9
1.75	50.2	34.86	6.29	2.99	60.91
1.8	51.6	35.83	6.62	3.14	65.89
1.85	53	36.81	6.96	3.3	71.17
1.9	54.5	37.85	7.33	3.47	77.07
1.95	55.9	38.82	7.68	3.63	82.82
2	**57.3**	**39.79**	**8.04**	**3.79**	**88.86**
2.05	58.8	40.83	8.44	3.98	95.72
2.1	60.2	41.81	8.81	4.15	102.32
2.15	61.6	42.78	9.2	4.33	109.33
2.2	63.1	43.82	9.61	4.52	116.98
2.25	64.5	44.79	10.01	4.7	124.54
2.3	65.9	45.76	10.42	4.89	132.45
2.35	67.4	46.81	10.86	5.09	141.21
2.4	**68.8**	**47.78**	**11.29**	**5.28**	**149.84**

C	80	90	100	110	120	130	140	145	150
Fq	.80	.90	1.00	1.10	1.20	1.30	1.40	1.45	1.50
Fhc	1.511	1.215	1.000	0.838	0.713	0.615	0.536	0.503	0.472
Fhm	1.498	1.210	1.000	0.842	0.719	0.622	0.544	0.510	0.480

Multiplying factors fq, fhc and fhm for dischange, head loss and modified head loss respectively wrt C values.

TABLE: HW 25 DISCHARGE AND LOSS OF HEAD AND POWER IN PIPES

(Hazen-Williams Formula, C=100)
(Modified Hazen-Williams Formula, CR=1)

Diameter of pipe 700 mm **Area 0.3848 sqm**

Velocity V, m/s	Flow ML/d	Flow kL/min	Head Loss Hfc,m/km	Head Loss Hfm,m/km	kW Wasted/km 60% efficiency
1	2	3	4	5	6
0.4	13.3	9.2	0.37	0.19	0.95
0.45	15	10.4	0.47	0.23	1.36
0.5	**16.6**	**11.5**	**0.56**	**0.28**	**1.79**
0.55	18.3	12.7	0.68	0.34	2.4
0.6	20	13.9	0.8	0.4	3.09
0.65	21.6	15	0.92	0.45	3.83
0.7	23.3	16.2	1.06	0.52	4.77
0.75	24.9	17.3	1.2	0.59	5.77
0.8	26.6	18.5	1.36	0.66	6.99
0.85	28.3	19.7	1.52	0.74	8.32
0.9	29.9	20.8	1.69	0.82	9.76
0.95	31.6	21.9	1.85	0.9	11.25
1	**33.3**	**23.1**	**2.05**	**0.99**	**13.15**
1.05	34.9	24.2	2.23	1.08	14.99
1.1	36.6	25.4	2.44	1.18	17.22
1.15	38.2	26.5	2.64	1.27	19.43
1.2	39.9	27.7	2.87	1.38	22.08
1.25	41.6	28.9	3.1	1.49	24.89
1.3	43.2	30	3.32	1.59	27.67
1.35	44.9	31.2	3.57	1.71	30.94
1.4	46.6	32.4	3.83	1.83	34.47
1.45	48.2	33.5	4.08	1.95	37.97
1.5	**49.9**	**34.7**	**4.35**	**2.07**	**41.93**
1.55	51.5	35.8	4.61	2.19	45.84
1.6	53.2	36.9	4.87	2.32	49.92
1.65	54.9	38.1	5.17	2.46	54.72
1.7	56.5	39.2	5.45	2.59	59.34
1.75	58.2	40.4	5.77	2.73	64.75
1.8	59.9	41.6	6.09	2.88	70.37
1.85	61.5	42.7	6.39	3.02	75.79
1.9	63.2	43.9	6.72	3.17	81.95
1.95	64.8	45	7.04	3.32	88
2	**66.5**	**46.2**	**7.39**	**3.48**	**94.84**
2.05	68.2	47.4	7.75	3.65	102.04
2.1	69.8	48.5	8.09	3.8	108.99
2.15	71.5	49.7	8.46	3.97	116.8
2.2	73.2	50.8	8.81	4.13	124.32
2.25	74.8	51.9	9.17	4.3	132.2
2.3	76.5	53.1	9.57	4.48	141.16
2.35	78.1	54.2	9.94	4.65	149.65
2.4	**79.8**	**55.4**	**10.35**	**4.84**	**159.28**

C	80	90	100	110	120	130	140	145	150
Fq	.80	.90	1.00	1.10	1.20	1.30	1.40	1.45	1.50
Fhc	1.511	1.215	1.000	0.838	0.713	0.615	0.536	0.503	0.472
Fhm	1.498	1.210	1.000	0.842	0.719	0.622	0.544	0.510	0.480

Multiplying factors fq, fhc and fhm for dischange, head loss and modified head loss respectively wrt C values.

TABLE: HW 26 DISCHARGE AND LOSS OF HEAD AND POWER IN PIPES
(Hazen-Williams Formula, C=100)
(Modified Hazen-Williams Formula, CR=1)

Diameter of pipe 750 mm **Area 0.4418 sqm**

Velocity V, m/s	*Flow ML/d*	*Flow kL/min*	*Head Loss Hfc,m/km*	*Head Loss Hfm,m/km*	*kW Wasted/km 60% efficiency*
1	*2*	*3*	*4*	*5*	*6*
0.4	15.3	10.6	0.35	0.17	1.03
0.45	17.2	11.9	0.43	0.21	1.42
0.5	**19.1**	**13.3**	**0.53**	**0.26**	**1.96**
0.55	21	14.6	0.63	0.31	2.56
0.6	22.9	15.9	0.73	0.36	3.22
0.65	24.8	17.2	0.85	0.42	4.06
0.7	26.7	18.5	0.97	0.48	4.98
0.75	28.6	19.9	1.11	0.54	6.14
0.8	30.5	21.2	1.25	0.61	7.36
0.85	32.4	22.5	1.39	0.68	8.69
0.9	34.4	23.9	1.56	0.76	10.36
0.95	36.3	25.2	1.72	0.83	12.04
1	**38.2**	**26.5**	**1.89**	**0.91**	**13.91**
1.05	40.1	27.8	2.06	1	15.91
1.1	42	29.2	2.26	1.09	18.33
1.15	43.9	30.5	2.45	1.18	20.76
1.2	45.8	31.8	2.64	1.27	23.32
1.25	47.7	33.1	2.85	1.37	26.2
1.3	49.6	34.4	3.06	1.46	29.24
1.35	51.5	35.8	3.29	1.57	32.72
1.4	53.4	37.1	3.52	1.68	36.28
1.45	55.3	38.4	3.75	1.79	40
1.5	**57.3**	**39.8**	**4.01**	**1.91**	**44.33**
1.55	59.2	41.1	4.25	2.02	48.52
1.6	61.1	42.4	4.51	2.14	53.12
1.65	63	43.8	4.79	2.27	58.28
1.7	64.9	45.1	5.05	2.39	63.27
1.75	66.8	46.4	5.32	2.52	68.57
1.8	68.7	47.7	5.6	2.65	74.2
1.85	70.6	49	5.89	2.78	80.17
1.9	72.5	50.3	6.18	2.91	86.35
1.95	74.4	51.7	6.51	3.06	93.49
2	**76.3**	**53**	**6.81**	**3.2**	**100.26**
2.05	78.3	54.4	7.15	3.36	108.04
2.1	80.2	55.7	7.47	3.5	115.58
2.15	82.1	57	7.79	3.65	123.34
2.2	84	58.3	8.13	3.81	131.66
2.25	85.9	59.7	8.49	3.97	140.79
2.3	87.8	61	8.84	4.13	149.79
2.35	89.7	62.3	9.19	4.29	159.04
2.4	**91.6**	**63.6**	**9.55**	**4.46**	**168.72**

C	80	90	100	110	120	130	140	145	150
Fq	.80	.90	1.00	1.10	1.20	1.30	1.40	1.45	1.50
Fhc	1.511	1.215	1.000	0.838	0.713	0.615	0.536	0.503	0.472
Fhm	1.498	1.210	1.000	0.842	0.719	0.622	0.544	0.510	0.480

Multiplying factors fq, fhc and fhm for dischange, head loss and modified head loss respectively wrt C values.

TABLE: HW 27 DISCHARGE AND LOSS OF HEAD AND POWER IN PIPES
(Hazen-Williams Formula, C=100)
(Modified Hazen-Williams Formula, CR=1)

Diameter of pipe 800 mm **Area 0.5027 sqm**

Velocity V, m/s	Flow ML/d	Flow kL/min	Head Loss Hfc,m/km	Head Loss Hfm,m/km	kW Wasted/km 60% efficiency
1	2	3	4	5	6
0.4	17.4	12.1	0.32	0.16	1.08
0.45	19.5	13.5	0.4	0.2	1.5
0.5	**21.7**	**15.1**	**0.49**	**0.24**	**2.06**
0.55	23.9	16.6	0.58	0.29	2.67
0.6	26.1	18.1	0.68	0.34	3.42
0.65	28.2	19.6	0.79	0.39	4.3
0.7	30.4	21.1	0.9	0.44	5.28
0.75	32.6	22.6	1.03	0.5	6.47
0.8	34.7	24.1	1.16	0.56	7.77
0.85	36.9	25.6	1.29	0.63	9.17
0.9	39.1	27.2	1.45	0.7	10.96
0.95	41.3	28.7	1.6	0.77	12.76
1	**43.4**	**30.1**	**1.74**	**0.84**	**14.55**
1.05	45.6	31.7	1.92	0.93	16.91
1.1	47.8	33.2	2.09	1.01	19.27
1.15	49.9	34.7	2.27	1.09	21.88
1.2	52.1	36.2	2.46	1.18	24.74
1.25	54.3	37.7	2.65	1.27	27.75
1.3	56.5	39.2	2.85	1.36	31.03
1.35	58.6	40.7	3.05	1.46	34.48
1.4	60.8	42.2	3.26	1.55	38.21
1.45	63	43.8	3.49	1.66	42.46
1.5	**65.1**	**45.2**	**3.7**	**1.76**	**46.46**
1.55	67.3	46.7	3.94	1.87	51.11
1.6	69.5	48.3	4.19	1.99	56.22
1.65	71.7	49.8	4.43	2.1	61.28
1.7	73.8	51.3	4.68	2.21	66.69
1.75	76	52.8	4.94	2.33	72.45
1.8	78.2	54.3	5.2	2.45	78.43
1.85	80.3	55.8	5.47	2.58	84.79
1.9	82.5	57.3	5.75	2.7	91.52
1.95	84.7	58.8	6.03	2.83	98.49
2	**86.9**	**60.3**	**6.32**	**2.97**	**105.86**
2.05	89	61.8	6.61	3.1	113.47
2.1	91.2	63.3	6.91	3.24	121.5
2.15	93.4	64.9	7.24	3.39	130.52
2.2	95.5	66.3	7.53	3.52	138.68
2.25	97.7	67.8	7.85	3.67	147.84
2.3	99.9	69.4	8.2	3.83	158.08
2.35	102.1	70.9	8.53	3.98	167.99
2.4	**104.2**	**72.4**	**8.86**	**4.13**	**178.18**

C	80	90	100	110	120	130	140	145	150
Fq	.80	.90	1.00	1.10	1.20	1.30	1.40	1.45	1.50
Fhc	1.511	1.215	1.000	0.838	0.713	0.615	0.536	0.503	0.472
Fhm	1.498	1.210	1.000	0.842	0.719	0.622	0.544	0.510	0.480

Multiplying factors fq, fhc and fhm for dischange, head loss and modified head loss respectively wrt C values.

TABLE: HW 28 DISCHARGE AND LOSS OF HEAD AND POWER IN PIPES
(Hazen-Williams Formula, C=100)
(Modified Hazen-Williams Formula, CR=1)

Diameter of pipe 850 mm **Area 0.5674 sqm**

Velocity V, m/s	*Flow ML/d*	*Flow kL/min*	*Head Loss Hfc,m/km*	*Head Loss Hfm,m/km*	*kW Wasted/km 60% efficiency*
1	*2*	*3*	*4*	*5*	*6*
0.4	19.6	13.6	0.3	0.15	1.13
0.45	22.1	15.3	0.37	0.19	1.57
0.5	**24.5**	**17**	**0.45**	**0.22**	**2.13**
0.55	27	18.8	0.54	0.27	2.82
0.6	29.4	20.4	0.63	0.31	3.57
0.65	31.9	22.2	0.74	0.36	4.56
0.7	34.3	23.8	0.84	0.41	5.55
0.75	36.8	25.6	0.96	0.47	6.83
0.8	39.2	27.2	1.08	0.52	8.16
0.85	41.7	29	1.21	0.59	9.75
0.9	44.1	30.6	1.34	0.65	11.39
0.95	46.6	32.4	1.49	0.72	13.41
1	**49**	**34**	**1.63**	**0.79**	**15.39**
1.05	51.5	35.8	1.79	0.86	17.8
1.1	53.9	37.4	1.94	0.93	20.15
1.15	56.4	39.2	2.12	1.02	23.08
1.2	58.8	40.8	2.28	1.09	25.84
1.25	61.3	42.6	2.47	1.18	29.23
1.3	63.7	44.2	2.65	1.26	32.54
1.35	66.2	46	2.85	1.36	36.42
1.4	68.6	47.6	3.03	1.44	40.06
1.45	71.1	49.4	3.25	1.54	44.6
1.5	**73.5**	**51**	**3.45**	**1.64**	**48.88**
1.55	76	52.8	3.68	1.74	53.97
1.6	78.4	54.4	3.89	1.84	58.78
1.65	80.9	56.2	4.13	1.95	64.47
1.7	83.3	57.8	4.35	2.05	69.84
1.75	85.8	59.6	4.6	2.17	76.16
1.8	88.3	61.3	4.85	2.28	82.58
1.85	90.7	63	5.1	2.4	89.25
1.9	93.2	64.7	5.36	2.52	96.33
1.95	95.6	66.4	5.62	2.64	103.66
2	**98.1**	**68.1**	**5.89**	**2.76**	**111.42**
2.05	100.5	69.8	6.17	2.89	119.63
2.1	103	71.5	6.45	3.02	128.1
2.15	105.4	73.2	6.73	3.15	136.84
2.2	107.9	74.9	7.03	3.28	146.26
2.25	110.3	76.6	7.32	3.42	155.75
2.3	112.8	78.3	7.63	3.56	165.95
2.35	115.2	80	7.94	3.7	176.44
2.4	**117.7**	**81.7**	**8.25**	**3.84**	**187.23**

C	80	90	100	110	120	130	140	145	150
Fq	.80	.90	1.00	1.10	1.20	1.30	1.40	1.45	1.50
Fhc	1.511	1.215	1.000	0.838	0.713	0.615	0.536	0.503	0.472
Fhm	1.498	1.210	1.000	0.842	0.719	0.622	0.544	0.510	0.480

Multiplying factors fq, fhc and fhm for dischange, head loss and modified head loss respectively wrt C values.

TABLE: HW 29 DISCHARGE AND LOSS OF HEAD AND POWER IN PIPES
(Hazen-Williams Formula, C=100)
(Modified Hazen-Williams Formula, CR=1)

Diameter of pipe 900 mm **Area 0.6362 sqm**

Velocity V, m/s	*Flow ML/d*	*Flow kL/min*	*Head Loss Hfc,m/km*	*Head Loss Hfm,m/km*	*kW Wasted/km 60% efficiency*
1	2	3	4	5	6
0.4	22	15.3	0.28	0.14	1.19
0.45	24.7	17.2	0.35	0.17	1.67
0.5	**27.5**	**19.1**	**0.42**	**0.21**	**2.23**
0.55	30.2	21	0.5	0.25	2.92
0.6	33	22.9	0.59	0.29	3.75
0.65	35.7	24.8	0.69	0.34	4.75
0.7	38.5	26.7	0.79	0.39	5.86
0.75	41.2	28.6	0.89	0.44	7.07
0.8	44	30.6	1.01	0.49	8.59
0.85	46.7	32.4	1.13	0.55	10.17
0.9	49.5	34.4	1.26	0.61	12.04
0.95	52.2	36.3	1.39	0.67	14.02
1	**55**	**38.2**	**1.53**	**0.74**	**16.24**
1.05	57.7	40.1	1.67	0.8	18.6
1.1	60.5	42	1.82	0.87	21.23
1.15	63.2	43.9	1.98	0.95	24.15
1.2	66	45.8	2.14	1.02	27.23
1.25	68.7	47.7	2.31	1.1	30.61
1.3	71.5	49.7	2.49	1.19	34.38
1.35	74.2	51.5	2.66	1.27	38.05
1.4	77	53.5	2.85	1.36	42.35
1.45	79.7	55.3	3.03	1.44	46.54
1.5	**82.4**	**57.2**	**3.23**	**1.53**	**51.32**
1.55	85.2	59.2	3.44	1.63	56.57
1.6	87.9	61	3.64	1.72	61.68
1.65	90.7	63	3.86	1.82	67.55
1.7	93.4	64.9	4.08	1.92	73.55
1.75	96.2	66.8	4.3	2.03	79.79
1.8	98.9	68.7	4.53	2.13	86.45
1.85	101.7	70.6	4.77	2.24	93.55
1.9	104.4	72.5	5.01	2.35	100.9
1.95	107.2	74.4	5.25	2.46	108.5
2	**109.9**	**76.3**	**5.5**	**2.58**	**116.57**
2.05	112.7	78.3	5.77	2.7	125.5
2.1	115.4	80.1	6.02	2.81	133.95
2.15	118.2	82.1	6.3	2.94	143.68
2.2	120.9	84	6.58	3.07	153.53
2.25	123.7	85.9	6.86	3.19	163.69
2.3	126.4	87.8	7.14	3.32	174.14
2.35	129.2	89.7	7.43	3.45	185.13
2.4	**131.9**	**91.6**	**7.72**	**3.59**	**196.43**

C	80	90	100	110	120	130	140	145	150
Fq	.80	.90	1.00	1.10	1.20	1.30	1.40	1.45	1.50
Fhc	1.511	1.215	1.000	0.838	0.713	0.615	0.536	0.503	0.472
Fhm	1.498	1.210	1.000	0.842	0.719	0.622	0.544	0.510	0.480

Multiplying factors fq, fhc and fhm for dischange, head loss and modified head loss respectively wrt C values.

TABLE: HW 30 DISCHARGE AND LOSS OF HEAD AND POWER IN PIPES
(Hazen-Williams Formula, C=100)
(Modified Hazen-Williams Formula, CR=1)

Diameter of pipe 950 mm **Area 0.7088 sqm**

Velocity V, m/s	*Flow ML/d*	*Flow kL/min*	*Head Loss Hfc,m/km*	*Head Loss Hfm,m/km*	*kW Wasted/km 60% efficiency*
1	*2*	*3*	*4*	*5*	*6*
0.4	24.5	17	0.26	0.13	1.23
0.45	27.6	19.2	0.33	0.16	1.76
0.5	**30.6**	**21.3**	**0.4**	**0.2**	**2.37**
0.55	33.7	23.4	0.47	0.23	3.06
0.6	36.7	25.5	0.56	0.27	3.97
0.65	39.8	27.6	0.64	0.32	4.91
0.7	42.9	29.8	0.74	0.36	6.13
0.75	45.9	31.9	0.84	0.41	7.44
0.8	49	34	0.95	0.46	8.97
0.85	52.1	36.2	1.06	0.52	10.66
0.9	55.1	38.3	1.18	0.57	12.55
0.95	58.2	40.4	1.3	0.63	14.59
1	**61.2**	**42.5**	**1.43**	**0.69**	**16.88**
1.05	64.3	44.7	1.57	0.75	19.49
1.1	67.4	46.8	1.71	0.82	22.23
1.15	70.4	48.9	1.86	0.89	25.27
1.2	73.5	51	2.01	0.96	28.48
1.25	76.6	53.2	2.17	1.03	32.07
1.3	79.6	55.3	2.33	1.11	35.79
1.35	82.7	57.4	2.5	1.19	39.86
1.4	85.7	59.5	2.67	1.27	44.13
1.45	88.8	61.7	2.85	1.35	48.85
1.5	**91.9**	**63.8**	**3.04**	**1.44**	**53.88**
1.55	94.9	65.9	3.22	1.52	58.94
1.6	98	68.1	3.43	1.62	64.88
1.65	101.1	70.2	3.63	1.71	70.79
1.7	104.1	72.3	3.83	1.8	76.92
1.75	107.2	74.4	4.04	1.9	83.49
1.8	110.2	76.5	4.25	2	90.31
1.85	113.3	78.7	4.48	2.1	97.94
1.9	116.4	80.8	4.7	2.2	105.49
1.95	119.4	82.9	4.93	2.31	113.53
2	**122.5**	**85.1**	**5.18**	**2.42**	**122.45**
2.05	125.5	87.2	5.42	2.53	131.28
2.1	128.6	89.3	5.66	2.64	140.4
2.15	131.7	91.5	5.92	2.76	150.47
2.2	134.7	93.5	6.16	2.87	159.99
2.25	137.8	95.7	6.44	2.99	171.2
2.3	140.9	97.8	6.7	3.11	182.02
2.35	143.9	99.9	6.97	3.24	193.42
2.4	**147**	**102.1**	**7.26**	**3.37**	**205.9**

C	80	90	100	110	120	130	140	145	150
Fq	.80	.90	1.00	1.10	1.20	1.30	1.40	1.45	1.50
Fhc	1.511	1.215	1.000	0.838	0.713	0.615	0.536	0.503	0.472
Fhm	1.498	1.210	1.000	0.842	0.719	0.622	0.544	0.510	0.480

Multiplying factors fq, fhc and fhm for dischange, head loss and modified head loss respectively wrt C values.

TABLE: HW 31 DISCHARGE AND LOSS OF HEAD AND POWER IN PIPES
(Hazen-Williams Formula, C=100)
(Modified Hazen-Williams Formula, CR=1)

Diameter of pipe 1000 mm **Area 0.7854 sqm**

Velocity V, m/s	*Flow ML/d*	*Flow kL/min*	*Head Loss Hfc,m/km*	*Head Loss Hfm,m/km*	*kW Wasted/km 60% efficiency*
1	*2*	*3*	*4*	*5*	*6*
0.4	27.1	18.8	0.25	0.12	1.31
0.45	30.5	21.2	0.31	0.15	1.83
0.5	**33.9**	**23.5**	**0.37**	**0.18**	**2.42**
0.55	37.3	25.9	0.45	0.22	3.24
0.6	40.7	28.3	0.52	0.26	4.09
0.65	44.1	30.6	0.61	0.3	5.19
0.7	47.5	33	0.7	0.34	6.42
0.75	50.9	35.3	0.79	0.38	7.75
0.8	54.3	37.7	0.89	0.43	9.32
0.85	57.7	40.1	1	0.48	11.14
0.9	61.1	42.4	1.11	0.54	13.07
0.95	64.5	44.8	1.23	0.59	15.31
1	**67.9**	**47.2**	**1.35**	**0.65**	**17.7**
1.05	71.3	49.5	1.48	0.71	20.35
1.1	74.6	51.8	1.61	0.77	23.17
1.15	78	54.2	1.75	0.84	26.35
1.2	81.4	56.5	1.89	0.9	29.66
1.25	84.8	58.9	2.04	0.97	33.38
1.3	88.2	61.3	2.2	1.04	37.46
1.35	91.6	63.6	2.35	1.12	41.52
1.4	95	66	2.52	1.19	46.2
1.45	98.4	68.3	2.68	1.27	50.85
1.5	**101.8**	**70.7**	**2.86**	**1.35**	**56.17**
1.55	105.2	73.1	3.04	1.44	61.73
1.6	108.6	75.4	3.22	1.52	67.44
1.65	112	77.8	3.42	1.61	73.91
1.7	115.4	80.1	3.61	1.7	80.32
1.75	118.8	82.5	3.81	1.79	87.31
1.8	122.1	84.8	4.01	1.88	94.46
1.85	125.5	87.2	4.22	1.98	102.22
1.9	128.9	89.5	4.43	2.07	110.13
1.95	132.3	91.9	4.65	2.17	118.7
2	**135.7**	**94.2**	**4.87**	**2.27**	**127.43**
2.05	139.1	96.6	5.1	2.38	136.85
2.1	142.5	99	5.34	2.49	146.85
2.15	145.9	101.3	5.57	2.59	156.73
2.2	149.3	103.7	5.82	2.71	167.65
2.25	152.7	106	6.06	2.82	178.43
2.3	156.1	108.4	6.31	2.93	190
2.35	159.5	110.8	6.58	3.05	202.52
2.4	**162.9**	**113.1**	**6.83**	**3.17**	**214.58**

C	80	90	100	110	120	130	140	145	150
Fq	.80	.90	1.00	1.10	1.20	1.30	1.40	1.45	1.50
Fhc	1.511	1.215	1.000	0.838	0.713	0.615	0.536	0.503	0.472
Fhm	1.498	1.210	1.000	0.842	0.719	0.622	0.544	0.510	0.480

Multiplying factors fq, fhc and fhm for dischange, head loss and modified head loss respectively wrt C values.

TABLE: HW 32 DISCHARGE AND LOSS OF HEAD AND POWER IN PIPES
(Hazen-Williams Formula, C=100)
(Modified Hazen-Williams Formula, CR=1)

Diameter of pipe 1100 mm **Area 0.9503 sqm**

Velocity V, m/s	*Flow ML/d*	*Flow kL/min*	*Head Loss Hfc,m/km*	*Head Loss Hfm,m/km*	*kW Wasted/km 60% efficiency*
1	2	*3*	4	5	6
0.4	32.8	22.8	0.22	0.11	1.39
0.45	36.9	25.6	0.27	0.14	1.92
0.5	**41.1**	**28.5**	**0.33**	**0.17**	**2.61**
0.55	45.2	31.4	0.4	0.2	3.49
0.6	49.3	34.2	0.47	0.23	4.47
0.65	53.4	37.1	0.54	0.27	5.57
0.7	57.5	39.9	0.62	0.3	6.87
0.75	61.6	42.8	0.71	0.34	8.44
0.8	65.7	45.6	0.8	0.39	10.13
0.85	69.8	48.5	0.9	0.43	12.13
0.9	73.9	51.3	0.99	0.48	14.11
0.95	78	54.2	1.1	0.53	16.56
1	**82.1**	**57**	**1.21**	**0.58**	**19.16**
1.05	86.2	59.9	1.32	0.63	21.96
1.1	90.3	62.7	1.44	0.69	25.08
1.15	94.4	65.6	1.57	0.75	28.61
1.2	98.5	68.4	1.69	0.81	32.11
1.25	102.6	71.3	1.83	0.87	36.24
1.3	106.7	74.1	1.96	0.93	40.34
1.35	110.8	76.9	2.1	1	44.86
1.4	115	79.9	2.26	1.07	50.16
1.45	119.1	82.7	2.4	1.14	55.13
1.5	**123.2**	**85.6**	**2.56**	**1.21**	**60.87**
1.55	127.3	88.4	2.72	1.28	66.79
1.6	131.4	91.3	2.89	1.36	73.29
1.65	135.5	94.1	3.05	1.43	79.72
1.7	139.6	96.9	3.23	1.51	86.94
1.75	143.7	99.8	3.41	1.6	94.53
1.8	147.8	102.6	3.59	1.68	102.32
1.85	151.9	105.5	3.78	1.76	110.78
1.9	156	108.3	3.96	1.85	119.13
1.95	160.1	111.2	4.16	1.94	128.5
2	**164.2**	**114**	**4.36**	**2.03**	**138.07**
2.05	168.3	116.9	4.57	2.12	148.4
2.1	172.4	119.7	4.77	2.22	158.6
2.15	176.5	122.6	4.99	2.32	169.94
2.2	180.6	125.4	5.2	2.41	181.13
2.25	184.7	128.3	5.42	2.51	193.16
2.3	188.9	131.2	5.65	2.62	205.91
2.35	193	134	5.88	2.72	218.87
2.4	**197.1**	**136.9**	**6.12**	**2.83**	**232.73**

C	80	90	100	110	120	130	140	145	150
Fq	.80	.90	1.00	1.10	1.20	1.30	1.40	1.45	1.50
Fhc	1.511	1.215	1.000	0.838	0.713	0.615	0.536	0.503	0.472
Fhm	1.498	1.210	1.000	0.842	0.719	0.622	0.544	0.510	0.480

Multiplying factors fq, fhc and fhm for dischange, head loss and modified head loss respectively wrt C values.

TABLE: HW 33 DISCHARGE AND LOSS OF HEAD AND POWER IN PIPES
(Hazen-Williams Formula, C=100)
(Modified Hazen-Williams Formula, CR=1)

Diameter of pipe 1200 mm **Area 1.1310 sqm**

Velocity V, m/s	*Flow ML/d*	*Flow kL/min*	*Head Loss Hfc,m/km*	*Head Loss Hfm,m/km*	*kW Wasted/km 60% efficiency*
1	*2*	*3*	*4*	*5*	*6*
0.4	39.1	27.2	0.2	0.1	1.51
0.45	44	30.6	0.25	0.12	2.13
0.5	**48.9**	**34**	**0.3**	**0.15**	**2.83**
0.55	53.7	37.3	0.36	0.18	3.73
0.6	58.6	40.7	0.42	0.21	4.75
0.65	63.5	44.1	0.49	0.24	6
0.7	68.4	47.5	0.56	0.27	7.39
0.75	73.3	50.9	0.64	0.31	9.05
0.8	78.2	54.3	0.72	0.35	10.86
0.85	83.1	57.7	0.81	0.39	12.98
0.9	87.9	61	0.9	0.43	15.25
0.95	92.8	64.4	0.99	0.48	17.71
1	**97.7**	**67.8**	**1.09**	**0.52**	**20.53**
1.05	102.6	71.3	1.2	0.57	23.77
1.1	107.5	74.7	1.3	0.62	26.98
1.15	112.4	78.1	1.42	0.67	30.81
1.2	117.3	81.5	1.53	0.73	34.64
1.25	122.1	84.8	1.65	0.78	38.87
1.3	127	88.2	1.77	0.84	43.37
1.35	131.9	91.6	1.9	0.9	48.34
1.4	136.8	95	2.04	0.96	53.83
1.45	141.7	98.4	2.17	1.02	59.31
1.5	**146.6**	**101.8**	**2.31**	**1.09**	**65.32**
1.55	151.5	105.2	2.46	1.16	71.89
1.6	156.3	108.5	2.6	1.22	78.36
1.65	161.2	111.9	2.76	1.29	85.79
1.7	166.1	115.3	2.91	1.36	93.2
1.75	171	118.8	3.08	1.44	101.64
1.8	175.9	122.2	3.24	1.52	109.98
1.85	180.8	125.6	3.41	1.59	118.97
1.9	185.7	129	3.59	1.67	128.64
1.95	190.6	132.4	3.76	1.75	138.28
2	**195.4**	**135.7**	**3.94**	**1.83**	**148.52**
2.05	200.3	139.1	4.12	1.92	159.19
2.1	205.2	142.5	4.31	2	170.6
2.15	210.1	145.9	4.5	2.09	182.38
2.2	215	149.3	4.7	2.18	194.92
2.25	219.9	152.7	4.9	2.27	207.84
2.3	224.8	156.1	5.11	2.36	221.58
2.35	229.6	159.4	5.31	2.45	235.12
2.4	**234.5**	**162.8**	**5.52**	**2.55**	**249.63**

C	80	90	100	110	120	130	140	145	150
Fq	.80	.90	1.00	1.10	1.20	1.30	1.40	1.45	1.50
Fhc	1.511	1.215	1.000	0.838	0.713	0.615	0.536	0.503	0.472
Fhm	1.498	1.210	1.000	0.842	0.719	0.622	0.544	0.510	0.480

Multiplying factors fq, fhc and fhm for dischange, head loss and modified head loss respectively wrt C values.

TABLE: HW 34 DISCHARGE AND LOSS OF HEAD AND POWER IN PIPES
(Hazen-Williams Formula, C=100)
(Modified Hazen-Williams Formula, CR=1)

Diameter of pipe 1300 mm **Area 1.3273 sqm**

Velocity V, m/s	*Flow ML/d*	*Flow kL/min*	*Head Loss Hfc,m/km*	*Head Loss Hfm,m/km*	*kW Wasted/km 60% efficiency*
1	2	3	4	5	6
0.4	45.9	31.9	0.18	0.09	1.60
0.45	51.6	35.8	0.23	0.11	2.29
0.5	**57.3**	**39.8**	**0.28**	**0.14**	**3.10**
0.55	63.1	43.8	0.33	0.16	4.02
0.6	68.8	47.8	0.39	0.19	5.18
0.65	74.5	51.7	0.45	0.22	6.46
0.7	80.3	55.8	0.51	0.25	7.91
0.75	86	59.7	0.58	0.28	9.62
0.8	91.7	63.7	0.66	0.32	11.68
0.85	97.5	67.7	0.74	0.35	13.92
0.9	103.2	71.7	0.82	0.39	16.33
0.95	108.9	75.6	0.9	0.43	18.90
1	**114.7**	**79.7**	**1**	**0.48**	**22.14**
1.05	120.4	83.6	1.09	0.52	25.31
1.1	126.2	87.6	1.19	0.56	28.96
1.15	131.9	91.6	1.29	0.61	32.82
1.2	137.6	95.6	1.39	0.66	36.91
1.25	143.4	99.6	1.5	0.71	41.50
1.3	149.1	103.5	1.62	0.76	46.58
1.35	154.8	107.5	1.73	0.82	51.66
1.4	160.6	111.5	1.85	0.87	57.30
1.45	166.3	115.5	1.98	0.93	63.53
1.5	**172**	**119.4**	**2.1**	**0.99**	**69.65**
1.55	177.8	123.5	2.24	1.05	76.84
1.6	183.5	127.4	2.37	1.11	83.87
1.65	189.2	131.4	2.51	1.18	91.62
1.7	195	135.4	2.66	1.24	100.05
1.75	200.7	139.4	2.8	1.31	108.42
1.8	206.4	143.3	2.95	1.38	117.43
1.85	212.2	147.4	3.11	1.45	127.34
1.9	217.9	151.3	3.26	1.52	137.01
1.95	223.6	155.3	3.42	1.59	147.54
2	**229.4**	**159.3**	**3.59**	**1.67**	**158.86**
2.05	235.1	163.3	3.76	1.74	170.56
2.1	240.8	167.2	3.93	1.82	182.53
2.15	246.6	171.3	4.11	1.9	195.57
2.2	252.3	175.2	4.28	1.98	208.29
2.25	258	179.2	4.46	2.06	222.01
2.3	263.8	183.2	4.65	2.15	236.63
2.35	269.5	187.2	4.84	2.23	251.68
2.4	**275.2**	**191.1**	**5.03**	**2.32**	**267.01**

C	80	90	100	110	120	130	140	145	150
Fq	.80	.90	1.00	1.10	1.20	1.30	1.40	1.45	1.50
Fhc	1.511	1.215	1.000	0.838	0.713	0.615	0.536	0.503	0.472
Fhm	1.498	1.210	1.000	0.842	0.719	0.622	0.544	0.510	0.480

Multiplying factors fq, fhc and fhm for dischange, head loss and modified head loss respectively wrt C values.

TABLE: HW 35 DISCHARGE AND LOSS OF HEAD AND POWER IN PIPES
(Hazen-Williams Formula, C=100)
(Modified Hazen-Williams Formula, CR=1)

Diameter of pipe 1400 mm **Area 1.5394 sqm**

Velocity V, m/s	*Flow ML/d*	*Flow kL/min*	*Head Loss Hfc,m/km*	*Head Loss Hfm,m/km*	*kW Wasted/km 60% efficiency*
1	*2*	*3*	*4*	*5*	*6*
0.4	53.2	36.9	0.17	0.08	1.74
0.45	59.9	41.6	0.21	0.1	2.43
0.5	**66.5**	**46.2**	**0.25**	**0.12**	**3.21**
0.55	73.2	50.8	0.3	0.15	4.23
0.6	79.8	55.4	0.35	0.17	5.39
0.65	86.5	60.1	0.41	0.2	6.84
0.7	93.1	64.7	0.47	0.23	8.45
0.75	99.8	69.3	0.54	0.26	10.4
0.8	106.4	73.9	0.6	0.29	12.32
0.85	113.1	78.5	0.67	0.32	14.61
0.9	119.7	83.1	0.75	0.36	17.31
0.95	126.4	87.8	0.83	0.4	20.24
1	**133**	**92.4**	**0.91**	**0.44**	**23.36**
1.05	139.7	97	1	0.48	26.94
1.1	146.3	101.6	1.09	0.52	30.76
1.15	153	106.3	1.18	0.56	34.84
1.2	159.6	110.8	1.28	0.6	39.4
1.25	166.3	115.5	1.38	0.65	44.28
1.3	172.9	120.1	1.48	0.7	49.37
1.35	179.6	124.7	1.59	0.75	55.08
1.4	186.2	129.3	1.7	0.8	61.06
1.45	192.9	134	1.82	0.85	67.74
1.5	199.5	138.5	1.93	0.91	74.25
1.55	206.2	143.2	2.05	0.96	81.54
1.6	212.8	147.8	2.18	1.02	89.5
1.65	219.5	152.4	2.31	1.08	97.79
1.7	226.1	157	2.44	1.14	106.41
1.75	232.8	161.7	2.57	1.2	115.44
1.8	239.4	166.3	2.71	1.26	125.19
1.85	246.1	170.9	2.85	1.32	135.3
1.9	252.7	175.5	2.99	1.39	145.76
1.95	259.4	180.1	3.14	1.46	157.09
2	**266**	**184.7**	**3.29**	**1.52**	**168.8**
2.05	272.7	189.4	3.45	1.6	181.51
2.1	279.3	194	3.6	1.67	194
2.15	286	198.6	3.76	1.74	207.43
2.2	292.6	203.2	3.93	1.81	221.83
2.25	299.3	207.8	4.09	1.89	236.08
2.3	305.9	212.4	4.26	1.96	251.34
2.35	312.6	217.1	4.44	2.04	267.76
2.4	**319.2**	**221.7**	**4.61**	**2.12**	**283.9**

C	80	90	100	110	120	130	140	145	150
Fq	.80	.90	1.00	1.10	1.20	1.30	1.40	1.45	1.50
Fhc	1.511	1.215	1.000	0.838	0.713	0.615	0.536	0.503	0.472
Fhm	1.498	1.210	1.000	0.842	0.719	0.622	0.544	0.510	0.480

Multiplying factors fq, fhc and fhm for dischange, head loss and modified head loss respectively wrt C values.

TABLE: HW 36 DISCHARGE AND LOSS OF HEAD AND POWER IN PIPES (Hazen-Williams Formula, C=100) (Modified Hazen-Williams Formula, CR=1)

Diameter of pipe 1500 mm **Area 1.7672 sqm**

Velocity V, m/s	*Flow ML/d*	*Flow kL/min*	*Head Loss Hfc,m/km*	*Head Loss Hfm,m/km*	*kW Wasted/km 60% efficiency*
1	*2*	*3*	*4*	*5*	*6*
0.4	61.1	42.4	0.15	0.08	1.77
0.45	68.7	47.7	0.19	0.09	2.52
0.5	**76.3**	**53**	**0.23**	**0.11**	**3.39**
0.55	84	58.3	0.28	0.14	4.53
0.6	91.6	63.6	0.33	0.16	5.83
0.65	99.2	68.9	0.38	0.18	7.27
0.7	106.9	74.2	0.43	0.21	8.86
0.75	114.5	79.5	0.49	0.24	10.82
0.8	122.1	84.8	0.56	0.27	13.19
0.85	129.8	90.1	0.62	0.3	15.52
0.9	137.4	95.4	0.69	0.33	18.29
0.95	145.1	100.8	0.77	0.37	21.56
1	**152.7**	**106**	**0.84**	**0.4**	**24.73**
1.05	160.3	111.3	0.92	0.44	28.44
1.1	168	116.7	1	0.48	32.42
1.15	175.6	121.9	1.09	0.52	36.91
1.2	183.2	127.2	1.18	0.56	41.69
1.25	190.9	132.6	1.27	0.6	46.78
1.3	198.5	137.8	1.37	0.64	52.44
1.35	206.1	143.1	1.47	0.69	58.43
1.4	213.8	148.5	1.57	0.74	64.76
1.45	221.4	153.8	1.68	0.79	71.77
1.5	**229**	**159**	**1.78**	**0.83**	**78.62**
1.55	236.7	164.4	1.9	0.89	86.77
1.6	244.3	169.7	2.01	0.94	94.75
1.65	251.9	174.9	2.13	0.99	103.48
1.7	259.6	180.3	2.25	1.05	112.69
1.75	267.2	185.6	2.37	1.1	122.19
1.8	274.8	190.8	2.5	1.16	132.5
1.85	282.5	196.2	2.63	1.22	143.34
1.9	290.1	201.5	2.76	1.28	154.48
1.95	297.7	206.7	2.9	1.34	166.51
2	**305.4**	**212.1**	**3.04**	**1.41**	**179.11**
2.05	313	217.4	3.18	1.47	192.04
2.1	320.6	222.6	3.32	1.53	205.29
2.15	328.3	228	3.47	1.6	219.77
2.2	335.9	233.3	3.62	1.67	234.6
2.25	343.5	238.5	3.78	1.74	250.43
2.3	351.2	243.9	3.94	1.81	266.94
2.35	358.8	249.2	4.1	1.88	283.81
2.4	**366.4**	**254.4**	**4.26**	**1.95**	**301.04**

C	80	90	100	110	120	130	140	145	150
Fq	.80	.90	1.00	1.10	1.20	1.30	1.40	1.45	1.50
Fhc	1.511	1.215	1.000	0.838	0.713	0.615	0.536	0.503	0.472
Fhm	1.498	1.210	1.000	0.842	0.719	0.622	0.544	0.510	0.480

Multiplying factors fq, fhc and fhm for dischange, head loss and modified head loss respectively wrt C values.

TABLE: HW 37 DISCHARGE AND LOSS OF HEAD AND POWER IN PIPES
(Hazen-Williams Formula, C=100)
(Modified Hazen-Williams Formula, CR=1)

Diameter of pipe 1600 mm **Area 2.0106 sqm**

Velocity V, m/s	*Flow ML/d*	*Flow kL/min*	*Head Loss Hfc,m/km*	*Head Loss Hfm,m/km*	*kW Wasted/km 60% efficiency*
1	*2*	*3*	*4*	*5*	*6*
0.4	69.5	48.3	0.14	0.07	1.88
0.45	78.2	54.3	0.18	0.09	2.72
0.5	**86.9**	**60.3**	**0.22**	**0.11**	**3.69**
0.55	95.5	66.3	0.26	0.13	4.79
0.6	104.2	72.4	0.3	0.15	6.03
0.65	112.9	78.4	0.35	0.17	7.62
0.7	121.6	84.4	0.4	0.19	9.38
0.75	130.3	90.5	0.46	0.22	11.56
0.8	139	96.5	0.52	0.25	13.94
0.85	147.7	102.6	0.58	0.28	16.53
0.9	156.3	108.5	0.64	0.31	19.29
0.95	165	114.6	0.71	0.34	22.6
1	**173.7**	**120.6**	**0.78**	**0.37**	**26.13**
1.05	182.4	126.7	0.85	0.41	29.92
1.1	191.1	132.7	0.93	0.44	34.28
1.15	199.8	138.8	1.01	0.48	38.94
1.2	208.5	144.8	1.09	0.52	43.84
1.25	217.2	150.8	1.18	0.56	49.43
1.3	225.8	156.8	1.27	0.6	55.32
1.35	234.5	162.8	1.36	0.64	61.5
1.4	243.2	168.9	1.46	0.68	68.5
1.45	251.9	174.9	1.55	0.73	75.3
1.5	260.6	181	1.65	0.77	82.96
1.55	269.3	187	1.76	0.82	91.42
1.6	278	193.1	1.86	0.87	99.77
1.65	286.6	199	1.97	0.92	108.9
1.7	295.3	205.1	2.09	0.97	119.07
1.75	304	211.1	2.2	1.02	129.01
1.8	312.7	217.2	2.32	1.08	139.97
1.85	321.4	223.2	2.44	1.13	151.28
1.9	330.1	229.2	2.56	1.19	162.99
1.95	338.8	235.3	2.69	1.24	175.82
2	**347.4**	**241.3**	**2.82**	**1.3**	**189.02**
2.05	356.1	247.3	2.95	1.36	202.65
2.1	364.8	253.3	3.08	1.42	216.71
2.15	373.5	259.4	3.22	1.48	232.02
2.2	382.2	265.4	3.36	1.55	247.71
2.25	390.9	271.5	3.51	1.61	264.71
2.3	399.6	277.5	3.65	1.68	281.35
2.35	408.2	283.5	3.8	1.74	299.25
2.4	**416.9**	**289.5**	**3.95**	**1.81**	**317.65**

C	80	90	100	110	120	130	140	145	150
Fq	.80	.90	1.00	1.10	1.20	1.30	1.40	1.45	1.50
Fhc	1.511	1.215	1.000	0.838	0.713	0.615	0.536	0.503	0.472
Fhm	1.498	1.210	1.000	0.842	0.719	0.622	0.544	0.510	0.480

Multiplying factors fq, fhc and fhm for dischange, head loss and modified head loss respectively wrt C values.

TABLE: HW 38 DISCHARGE AND LOSS OF HEAD AND POWER IN PIPES
(Hazen-Williams Formula, C=100)
(Modified Hazen-Williams Formula, CR=1)

Diameter of pipe 1700 mm **Area 2.2698 sqm**

Velocity V, m/s	*Flow ML/d*	*Flow kL/min*	*Head Loss Hfc,m/km*	*Head Loss Hfm,m/km*	*kW Wasted/km 60% efficiency*
1	2	3	4	5	6
0.4	78.4	54.4	0.13	0.07	1.96
0.45	88.3	61.3	0.17	0.08	2.89
0.5	**98.1**	**68.1**	**0.2**	**0.1**	**3.78**
0.55	107.9	74.9	0.24	0.12	4.99
0.6	117.7	81.7	0.28	0.14	6.35
0.65	127.5	88.5	0.33	0.16	8.11
0.7	137.3	95.3	0.38	0.18	10.06
0.75	147.1	102.2	0.43	0.21	12.21
0.8	156.9	109	0.48	0.23	14.53
0.85	166.7	115.8	0.54	0.26	17.37
0.9	176.5	122.6	0.6	0.29	20.43
0.95	186.3	129.4	0.66	0.31	23.72
1	**196.1**	**136.2**	**0.73**	**0.35**	**27.62**
1.05	205.9	143	0.8	0.38	31.78
1.1	215.7	149.8	0.87	0.41	36.2
1.15	225.5	156.6	0.94	0.44	40.89
1.2	235.3	163.4	1.02	0.48	46.3
1.25	245.1	170.2	1.1	0.52	52.01
1.3	255	177.1	1.18	0.56	58.05
1.35	264.8	183.9	1.27	0.59	64.88
1.4	274.6	190.7	1.36	0.63	72.04
1.45	284.4	197.5	1.45	0.68	79.55
1.5	**294.2**	**204.3**	**1.54**	**0.72**	**87.4**
1.55	304	211.1	1.64	0.76	96.17
1.6	313.8	217.9	1.74	0.81	105.32
1.65	323.6	224.7	1.84	0.85	114.85
1.7	333.4	231.5	1.94	0.9	124.75
1.75	343.2	238.3	2.05	0.95	135.7
1.8	353	245.1	2.16	1	147.06
1.85	362.8	251.9	2.27	1.05	158.84
1.9	372.6	258.8	2.39	1.1	171.81
1.95	382.4	265.6	2.51	1.16	185.18
2	**392.2**	**272.4**	**2.63**	**1.21**	**199**
2.05	402	279.2	2.75	1.27	213.28
2.1	411.8	286	2.87	1.32	228.01
2.15	421.6	292.8	3	1.38	244
2.2	431.5	299.7	3.13	1.44	260.57
2.25	441.3	306.5	3.27	1.5	278.4
2.3	451.1	313.3	3.4	1.56	295.89
2.35	460.9	320.1	3.54	1.62	314.77
2.4	**470.7**	**326.9**	**3.68**	**1.68**	**334.16**

C	80	90	100	110	120	130	140	145	150
Fq	.80	.90	1.00	1.10	1.20	1.30	1.40	1.45	1.50
Fhc	1.511	1.215	1.000	0.838	0.713	0.615	0.536	0.503	0.472
Fhm	1.498	1.210	1.000	0.842	0.719	0.622	0.544	0.510	0.480

Multiplying factors fq, fhc and fhm for dischange, head loss and modified head loss respectively wrt C values.

TABLE: HW 39 DISCHARGE AND LOSS OF HEAD AND POWER IN PIPES
(Hazen-Williams Formula, C=100)
(Modified Hazen-Williams Formula, CR=1)

Diameter of pipe 1800 mm **Area 2.5447 sqm**

Velocity V, m/s	*Flow ML/d*	*Flow kL/min*	*Head Loss Hfc,m/km*	*Head Loss Hfm,m/km*	*kW Wasted/km 60% efficiency*
1	2	3	4	5	6
0.4	87.9	61	0.12	0.06	2.03
0.45	98.9	68.7	0.15	0.08	2.86
0.5	**109.9**	**76.3**	**0.19**	**0.09**	**4.03**
0.55	120.9	84	0.22	0.11	5.13
0.6	131.9	91.6	0.26	0.13	6.62
0.65	142.9	99.2	0.31	0.15	8.54
0.7	153.9	106.9	0.35	0.17	10.39
0.75	164.9	114.5	0.4	0.19	12.72
0.8	175.9	122.2	0.45	0.22	15.28
0.85	186.9	129.8	0.5	0.24	18.03
0.9	197.9	137.4	0.56	0.27	21.37
0.95	208.9	145.1	0.62	0.29	24.99
1	**219.9**	**152.7**	**0.68**	**0.32**	**28.84**
1.05	230.9	160.3	0.74	0.35	32.95
1.1	241.9	168	0.81	0.38	37.8
1.15	252.8	175.6	0.88	0.42	42.92
1.2	263.8	183.2	0.95	0.45	48.34
1.25	274.8	190.8	1.03	0.48	54.59
1.3	285.8	198.5	1.11	0.52	61.2
1.35	296.8	206.1	1.19	0.56	68.13
1.4	307.8	213.8	1.27	0.59	75.42
1.45	318.8	221.4	1.35	0.63	83.03
1.5	**329.8**	**229**	**1.44**	**0.67**	**91.6**
1.55	340.8	236.7	1.53	0.71	100.6
1.6	351.8	244.3	1.62	0.76	109.94
1.65	362.8	251.9	1.72	0.8	120.35
1.7	373.8	259.6	1.82	0.84	131.24
1.75	384.8	267.2	1.92	0.89	142.51
1.8	395.8	274.9	2.02	0.93	154.25
1.85	406.8	282.5	2.13	0.98	167.15
1.9	417.7	290.1	2.23	1.03	179.7
1.95	428.7	297.7	2.34	1.08	193.51
2	**439.7**	**305.3**	**2.45**	**1.13**	**207.77**
2.05	450.7	313	2.57	1.18	223.45
2.1	461.7	320.6	2.69	1.23	239.56
2.15	472.7	328.3	2.81	1.29	256.26
2.2	483.7	335.9	2.93	1.34	273.39
2.25	494.7	343.5	3.05	1.4	291.02
2.3	505.7	351.2	3.18	1.46	310.23
2.35	516.7	358.8	3.31	1.51	329.9
2.4	**527.7**	**366.5**	**3.44**	**1.57**	**350.21**

C	80	90	100	110	120	130	140	145	150
Fq	.80	.90	1.00	1.10	1.20	1.30	1.40	1.45	1.50
Fhc	1.511	1.215	1.000	0.838	0.713	0.615	0.536	0.503	0.472
Fhm	1.498	1.210	1.000	0.842	0.719	0.622	0.544	0.510	0.480

Multiplying factors fq, fhc and fhm for dischange, head loss and modified head loss respectively wrt C values.

TABLE: HW 40 DISCHARGE AND LOSS OF HEAD AND POWER IN PIPES
(Hazen-Williams Formula, C=100)
(Modified Hazen-Williams Formula, CR=1)

Diameter of pipe 1900 mm **Area 2.8353 sqm**

Velocity V, m/s	*Flow ML/d*	*Flow kL/min*	*Head Loss Hfc,m/km*	*Head Loss Hfm,m/km*	*kW Wasted/km 60% efficiency*
1	*2*	*3*	*4*	*5*	*6*
0.4	98	68.1	0.12	0.06	2.27
0.45	110.2	76.5	0.15	0.07	3.19
0.5	**122.5**	**85.1**	**0.18**	**0.09**	**4.26**
0.55	134.7	93.5	0.21	0.1	5.45
0.6	147	102.1	0.25	0.12	7.09
0.65	159.2	110.6	0.29	0.14	8.91
0.7	171.5	119.1	0.33	0.16	10.92
0.75	183.7	127.6	0.37	0.18	13.11
0.8	196	136.1	0.42	0.2	15.88
0.85	208.2	144.6	0.47	0.23	18.88
0.9	220.5	153.1	0.53	0.25	22.54
0.95	232.7	161.6	0.58	0.28	26.04
1	**245**	**170.1**	**0.64**	**0.3**	**30.24**
1.05	257.2	178.6	0.7	0.33	34.73
1.1	269.5	187.2	0.76	0.36	39.52
1.15	281.7	195.6	0.83	0.39	45.1
1.2	294	204.2	0.9	0.42	51.05
1.25	306.2	212.6	0.97	0.45	57.28
1.3	318.5	221.2	1.04	0.49	63.9
1.35	330.7	229.7	1.11	0.52	70.82
1.4	343	238.2	1.19	0.56	78.74
1.45	355.2	246.7	1.27	0.59	87.03
1.5	**367.5**	**255.2**	**1.35**	**0.63**	**95.7**
1.55	379.7	263.7	1.44	0.67	105.48
1.6	392	272.2	1.53	0.71	115.69
1.65	404.2	280.7	1.61	0.75	125.54
1.7	416.5	289.2	1.71	0.79	137.37
1.75	428.7	297.7	1.8	0.83	148.85
1.8	441	306.3	1.9	0.88	161.66
1.85	453.2	314.7	2	0.92	174.83
1.9	465.5	323.3	2.1	0.97	188.59
1.95	477.7	331.7	2.2	1.01	202.71
2	**489.9**	**340.2**	**2.31**	**1.06**	**218.3**
2.05	502.2	348.8	2.41	1.11	233.5
2.1	514.4	357.2	2.52	1.16	250.04
2.15	526.7	365.8	2.64	1.21	268.25
2.2	538.9	374.2	2.75	1.26	285.85
2.25	551.2	382.8	2.87	1.31	305.18
2.3	563.4	391.3	2.99	1.37	325
2.35	575.7	399.8	3.11	1.42	345.38
2.4	**587.9**	**408.3**	**3.23**	**1.47**	**366.34**

C	80	90	100	110	120	130	140	145	150
Fq	.80	.90	1.00	1.10	1.20	1.30	1.40	1.45	1.50
Fhc	1.511	1.215	1.000	0.838	0.713	0.615	0.536	0.503	0.472
Fhm	1.498	1.210	1.000	0.842	0.719	0.622	0.544	0.510	0.480

Multiplying factors fq, fhc and fhm for dischange, head loss and modified head loss respectively wrt C values.

TABLE: HW 41 DISCHARGE AND LOSS OF HEAD AND POWER IN PIPES
(Hazen-Williams Formula, C=100)
(Modified Hazen-Williams Formula, CR=1)

Diameter of pipe 2000 mm **Area 3.1416 sqm**

Velocity V, m/s	*Flow ML/d*	*Flow kL/min*	*Head Loss Hfc,m/km*	*Head Loss Hfm,m/km*	*kW Wasted/km 60% efficiency*
1	*2*	*3*	*4*	*5*	*6*
0.4	108.6	75.4	0.11	0.05	2.3
0.45	122.1	84.8	0.14	0.07	3.3
0.5	**135.7**	**94.2**	**0.17**	**0.08**	**4.45**
0.55	149.3	103.7	0.2	0.1	5.76
0.6	162.9	113.1	0.23	0.11	7.23
0.65	176.4	122.5	0.27	0.13	9.19
0.7	190	131.9	0.31	0.15	11.36
0.75	203.6	141.4	0.35	0.17	13.75
0.8	217.2	150.8	0.4	0.19	16.76
0.85	230.7	160.2	0.45	0.21	20.03
0.9	244.3	169.7	0.5	0.24	23.57
0.95	257.9	179.1	0.55	0.26	27.36
1	**271.4**	**188.5**	**0.6**	**0.28**	**31.42**
1.05	285	197.9	0.66	0.31	36.28
1.1	298.6	207.4	0.72	0.34	41.48
1.15	312.2	216.8	0.78	0.37	46.97
1.2	325.7	226.2	0.84	0.4	52.78
1.25	339.3	235.6	0.91	0.43	59.55
1.3	352.9	245.1	0.98	0.46	66.72
1.35	366.4	254.4	1.05	0.49	74.2
1.4	380	263.9	1.12	0.52	82.1
1.45	393.6	273.3	1.2	0.56	91.1
1.5	**407.2**	**282.8**	**1.28**	**0.59**	**100.55**
1.55	420.7	292.2	1.35	0.63	109.58
1.6	434.3	301.6	1.44	0.67	120.64
1.65	447.9	311	1.52	0.7	131.31
1.7	461.4	320.4	1.61	0.74	143.29
1.75	475	329.9	1.7	0.78	155.79
1.8	488.6	339.3	1.79	0.82	168.71
1.85	502.2	348.8	1.88	0.87	182.15
1.9	515.7	358.1	1.97	0.91	195.96
1.95	529.3	367.6	2.07	0.95	211.37
2	**542.9**	**377**	**2.17**	**1**	**227.25**
2.05	556.5	386.5	2.27	1.04	243.71
2.1	570	395.8	2.38	1.09	261.67
2.15	583.6	405.3	2.48	1.14	279.21
2.2	597.2	414.7	2.59	1.19	298.35
2.25	610.7	424.1	2.7	1.23	318.08
2.3	624.3	433.5	2.81	1.28	338.37
2.35	637.9	443	2.93	1.34	360.55
2.4	**651.5**	**452.4**	**3.04**	**1.39**	**382.03**

C	80	90	100	110	120	130	140	145	150
Fq	.80	.90	1.00	1.10	1.20	1.30	1.40	1.45	1.50
Fhc	1.511	1.215	1.000	0.838	0.713	0.615	0.536	0.503	0.472
Fhm	1.498	1.210	1.000	0.842	0.719	0.622	0.544	0.510	0.480

Multiplying factors fq, fhc and fhm for dischange, head loss and modified head loss respectively wrt C values.

SECTION II

Pipes Flowing Full Under Gravity
Tables for Discharge and Gradients Required

TABLE: G01 Velocity, Discharge and Gradients for different Values of Mannings 'n' for Gravity Flow with Pipe Running Full.

Diameter: 100 mm **Area : 7.854E-3 sqm**

Velocity	*Flow*		*Gradient : 1 : L*		
V, m/s	*ML/d*	*kL/min*	*n = 0.015*	*n = 0.013*	*n = 0.011*
1	*2*	*3*	*4*	*5*	*6*
0.4	0.27	0.19	203	271	378
0.425	0.29	0.2	180	240	335
0.45	0.31	0.22	160	214	299
0.475	0.32	0.22	144	192	268
0.5	**0.34**	**0.24**	**130**	**173**	**242**
0.525	0.36	0.25	118	157	219
0.55	0.37	0.26	107	143	200
0.575	0.39	0.27	98	131	183
0.6	0.41	0.28	90	120	168
0.625	0.42	0.29	83	111	155
0.65	0.44	0.31	77	102	143
0.675	0.46	0.32	71	95	133
0.7	0.48	0.33	66	88	123
0.725	0.49	0.34	62	82	115
0.75	0.51	0.35	58	77	107
0,775	0.53	0.37	54	72	101
0.8	0.54	0.38	51	68	94
0.825	0.56	0.39	48	64	89
0.85	0.58	0.4	45	60	84
0.875	0.59	0.41	42	57	79
0.9	0.61	0.42	40	53	75
0.925	0.63	0.44	38	51	71
0.95	0.64	0.44	36	48	67
0.975	0.66	0.46	34	46	64
1	**0.68**	**0.47**	**32**	**43**	**60**
1.025	0.7	0.49	31	41	58
1.05	0.71	0.49	29	39	55
1.075	0.73	0.51	28	37	52
1.1	0.75	0.52	27	36	50
1.125	0.76	0.53	26	34	48
1.15	0.78	0.54	25	33	46
1.175	0.8	0.56	24	31	44
1.2	0.81	0.56	23	30	42
1.225	0.83	0.58	22	29	40
1.25	0.85	0.59	21	28	39
1.275	0.87	0.6	20	27	37
1.3	0.88	0.61	19	26	36
1.325	0.9	0.63	19	25	34
1.35	0.92	0.64	18	24	33
1.375	0.93	0.65	17	23	32
1.4	0.95	0.66	17	22	31
1.425	0.97	0.67	16	21	30
1.45	0.98	0.68	15	21	29
1.475	1	0.69	15	20	28
1.5	**1.02**	**0.71**	**14**	**19**	**27**

TABLE: G02 Velocity, Discharge and Gradients for different Values of Mannings 'n' for Gravity Flow with Pipe Running Full.

Diameter: 150 mm **Area :1.7672E-2 sqm**

Velocity	*Flow*		*Gradient : 1 : L*		
V, m/s	*ML/d*	*kL/min*	*n = 0.015*	*n = 0.013*	*n = 0.011*
1	*2*	*3*	*4*	*5*	*6*
0.4	0.61	0.42	349	465	649
0.425	0.65	0.45	309	411	575
0.45	0.69	0.48	276	367	513
0.475	0.73	0.51	247	329	460
0.5	**0.76**	**0.53**	**223**	**297**	**415**
0.525	0.8	0.56	202	270	377
0.55	0.84	0.58	184	246	343
0.575	0.88	0.61	169	225	314
0.6	0.92	0.64	155	206	288
0.625	0.95	0.66	143	190	266
0.65	0.99	0.69	132	176	246
0.675	1.03	0.72	122	163	228
0.7	1.07	0.74	114	152	212
0.725	1.11	0.77	106	141	197
0.75	1.15	0.8	99	132	185
0.775	1.18	0.82	93	124	173
0.8	1.22	0.85	87	116	162
0.825	1.26	0.88	82	109	153
0.85	1.3	0.9	77	103	144
0.875	1.34	0.93	73	97	136
0.9	1.37	0.95	69	92	128
0.925	1.41	0.98	65	87	121
0.95	1.45	1.01	62	82	115
0.975	1.49	1.03	59	78	109
1	**1.53**	**1.06**	**56**	**74**	**104**
1.025	1.57	1.09	53	71	99
1.05	1.6	1.11	51	67	94
1.075	1.64	1.14	48	64	90
1.1	1.68	1.17	46	61	86
1.125	1.72	1.19	44	59	82
1.15	1.76	1.22	42	56	78
1.175	1.79	1.24	40	54	75
1.2	1.83	1.27	39	52	72
1.225	1.87	1.3	37	50	69
1.25	1.91	1.33	36	48	66
1.275	1.95	1.35	34	46	64
1.3	1.98	1.38	33	44	61
1.325	2.02	1.4	32	42	59
1.35	2.06	1.43	31	41	57
1.375	2.1	1.46	30	39	55
1.4	2.14	1.49	28	38	53
1.425	2.18	1.51	27	37	51
1.45	2.21	1.53	27	35	49
1.475	2.25	1.56	26	34	48
1.5	**2.29**	**1.59**	**25**	**33**	**46**

TABLE : G03 Velocity, Discharge and Gradients for different Values of Mannings 'n' for Gravity Flow with Pipe Running Full.

Diameter: 200 mm **Area : 3.1416E-2 sqm**

Velocity	*Flow*		*Gradient : 1 : L*		
V, m/s	*ML/d*	*kL/min*	*n = 0.015*	*n = 0.013*	*n = 0.011*
1	*2*	*3*	*4*	*5*	*6*
0.4	1.09	0.76	512	682	952
0.425	1.15	0.8	453	604	843
0.45	1.22	0.85	404	539	752
0.475	1.29	0.9	363	483	675
0.5	**1.36**	**0.94**	**328**	**436**	**609**
0.525	1.43	0.99	297	396	553
0.55	1.49	1.03	271	361	504
0.575	1.56	1.08	248	330	461
0.6	1.63	1.13	227	303	423
0.625	1.7	1.18	210	279	390
0.65	1.76	1.22	194	258	361
0.675	1.83	1.27	180	239	334
0.7	1.9	1.32	167	223	311
0.725	1.97	1.37	156	207	290
0.75	2.04	1.42	146	194	271
0.775	2.1	1.46	136	182	254
0.8	2.17	1.51	128	170	238
0.825	2.24	1.56	120	160	224
0.85	2.31	1.6	113	151	211
0.875	2.38	1.65	107	142	199
0.9	2.44	1.69	101	135	188
0.925	2.51	1.74	96	127	178
0.95	2.58	1.79	91	121	169
0.975	2.65	1.84	86	115	160
1	**2.71**	**1.88**	**82**	**109**	**152**
1.025	2.78	1.93	78	104	145
1.05	2.85	1.98	74	99	138
1.075	2.92	2.03	71	94	132
1.1	2.99	2.08	68	90	126
1.125	3.05	2.12	65	86	120
1.15	3.12	2.17	62	82	115
1.175	3.19	2.22	59	79	110
1.2	3.26	2.26	57	76	106
1.225	3.33	2.31	55	73	102
1.25	3.39	2.35	52	70	97
1.275	3.46	2.4	50	67	94
1.3	3.53	2.45	48	65	90
1.325	3.6	2.5	47	62	87
1.35	3.66	2.54	45	60	84
1.375	3.73	2.59	43	58	81
1.4	3.8	2.64	42	56	78
1.425	3.87	2.69	40	54	75
1.45	3.94	2.74	39	52	72
1.475	4	2.78	38	50	70
1.5	**4.07**	**2.83**	**36**	**48**	**68**

TABLE : G04 Velocity, Discharge and Gradients for different Values of Mannings 'n' for Gravity Flow with Pipe Running Full.

Diameter: 250 mm **Area : 4.9088E-2 sqm**

Velocity V, m/s	*Flow*		*Gradient : 1 : L*		
	ML/d	*kL/min*	*n = 0.015*	*n = 0.013*	*n = 0.011*
1	*2*	*3*	*4*	*5*	*6*
0.4	1.7	1.18	689	918	1282
0.425	1.8	1.25	610	813	1135
0.45	1.91	1.33	544	725	1013
0.475	2.01	1.4	489	651	909
0.5	**2.12**	**1.47**	**441**	**587**	**820**
0.525	2.23	1.55	400	533	744
0.55	2.33	1.62	364	485	678
0.575	2.44	1.69	333	444	620
0.6	2.54	1.76	306	408	570
0.625	2.65	1.84	282	376	525
0.65	2.76	1.92	261	348	485
0.675	2.86	1.99	242	322	450
0.7	2.97	2.06	225	300	419
0.725	3.07	2.13	210	279	390
0.75	3.18	2.21	196	261	365
0.775	3.29	2.28	184	244	341
0.8	3.39	2.35	172	229	320
0.825	3.5	2.43	162	216	301
0.85	3.61	2.51	153	203	284
0.875	3.71	2.58	144	192	268
0.9	3.82	2.65	136	181	253
0.925	3.92	2.72	129	172	240
0.95	4.03	2.8	122	163	227
0.975	4.14	2.88	116	154	216
1	**4.24**	**2.94**	**110**	**147**	**205**
1.025	4.35	3.02	105	140	195
1.05	4.45	3.09	100	133	186
1.075	4.56	3.17	95	127	177
1.1	4.67	3.24	91	121	169
1.125	4.77	3.31	87	116	162
1.15	4.88	3.39	83	111	155
1.175	4.98	3.46	80	106	149
1.2	5.09	3.53	77	102	142
1.225	5.2	3.61	73	98	137
1.25	5.3	3.68	71	94	131
1.275	5.41	3.76	68	90	126
1.3	5.51	3.83	65	87	121
1.325	5.62	3.9	63	84	117
1.35	5.73	3.98	60	81	113
1.375	5.83	4.05	58	78	108
1.4	5.94	4.13	56	75	105
1.425	6.04	4.19	54	72	101
1.45	6.15	4.27	52	70	98
1.475	6.26	4.35	51	67	94
1.5	**6.36**	**4.42**	**49**	**65**	**91**

TABLE : G05 Velocity, Discharge and Gradients for different Values of Mannings 'n' for Gravity Flow with pipe Running Full.

Diameter: 300 mm **Area : 7.0686E-2 sqm**

Velocity	*Flow*		*Gradient : 1 : L*		
V, m/s	*ML/d*	*kL/min*	*n = 0.015*	*n = 0.013*	*n = 0.011*
1	*2*	*3*	*4*	*5*	*6*
0.4	2.44	1.69	879	1170	1634
0.425	2.6	1.81	778	1037	1448
0.45	2.75	1.91	694	925	1291
0.475	2.9	2.01	623	830	1159
0.5	**3.05**	**2.12**	**562**	**749**	**1046**
0.525	3.21	2.23	510	679	949
0.55	3.36	2.33	465	619	865
0.575	3.51	2.44	425	566	791
0.6	3.66	2.54	391	520	726
0.625	3.82	2.65	360	479	669
0.65	3.97	2.76	333	443	619
0.675	4.12	2.86	309	411	574
0.7	4.28	2.97	287	382	534
0.725	4.43	3.08	267	356	498
0.75	4.58	3.18	250	333	465
0.775	4.73	3.28	234	312	435
0.8	4.89	3.4	220	293	409
0.825	5.04	3.5	207	275	384
0.85	5.19	3.6	195	259	362
0.875	5.34	3.71	184	245	342
0.9	5.5	3.82	174	231	323
0.925	5.65	3.92	164	219	306
0.95	5.8	4.03	156	207	290
0.975	5.95	4.13	148	197	275
1	**6.11**	**4.24**	**141**	**187**	**262**
1.025	6.26	4.35	134	178	249
1.05	6.41	4.45	128	170	237
1.075	6.57	4.56	122	162	226
1.1	6.72	4.67	116	155	216
1.125	6.87	4.77	111	148	207
1.15	7.02	4.88	106	142	198
1.175	7.18	4.99	102	136	189
1.2	7.33	5.09	98	130	182
1.225	7.48	5.19	94	125	174
1.25	7.63	5.3	90	120	167
1.275	7.79	5.41	86	115	161
1.3	7.94	5.51	83	111	155
1.325	8.09	5.62	80	107	149
1.35	8.24	5.72	77	103	143
1.375	8.4	5.83	74	99	138
1.4	8.55	5.94	72	96	133
1.425	8.7	6.04	69	92	129
1.45	8.86	6.15	67	89	124
1.475	9.01	6.26	65	86	120
1.5	**9.16**	**6.36**	**62**	**83**	**116**

TABLE: G06 Velocity, Discharge and Gradients for different Values of Mannings 'n' for Gravity Flow with Pipe Running Full.

Diameter: 350 mm **Area: 9.6211E-2 sqm**

Velocity V, m/s	*Flow*		*Gradient : 1 : L*		
	ML/d	*kL/min*	*n = 0.015*	*n = 0.013*	*n = 0.011*
1	*2*	*3*	*4*	*5*	*6*
0.4	3.33	2.31	1079	1437	2007
0.425	3.53	2.45	956	1273	1778
0.45	3.74	2.6	853	1136	1586
0.475	3.95	2.74	765	1019	1423
0.5	**4.16**	**2.89**	**691**	**920**	**1285**
0.525	4.36	3.03	626	834	1165
0.55	4.57	3.17	571	760	1062
0.575	4.78	3.32	522	695	971
0.6	4.99	3.47	480	639	892
0.625	5.2	3.61	442	589	822
0.65	5.4	3.75	409	544	760
0.675	5.61	3.9	379	505	705
0.7	5.82	4.04	352	469	655
0.725	6.03	4.19	328	437	611
0.75	6.23	4.33	307	409	571
0.775	6.44	4.47	287	383	535
0.8	6.65	4.62	270	359	502
0.825	6.86	4.76	254	338	472
0.85	7.07	4.91	239	318	445
0.875	7.27	5.05	226	300	419
0.9	7.48	5.19	213	284	397
0.925	7.69	5.34	202	269	375
0.95	7.9	5.49	191	255	356
0.975	8.11	5.63	182	242	338
1	**8.31**	**5.77**	**173**	**230**	**321**
1.025	8.52	5.92	164	219	306
1.05	8.73	6.06	157	209	291
1.075	8.94	6.21	149	199	278
1.1	9.14	6.35	143	190	265
1.125	9.35	6.49	136	182	254
1.15	9.56	6.64	131	174	243
1.175	9.77	6.78	125	167	233
1.2	9.98	6.93	120	160	223
1.225	10.18	7.07	115	153	214
1.25	10.39	7.22	111	147	206
1.275	10.6	7.36	106	141	198
1.3	10.81	7.51	102	136	190
1.325	11.01	7.65	98	131	183
1.35	11.22	7.79	95	126	176
1.375	11.43	7.94	91	122	170
1.4	11.64	8.08	88	117	164
1.425	11.85	8.23	85	113	158
1.45	12.05	8.37	82	109	153
1.475	12.26	8.51	79	106	148
1.5	**12.47**	**74.9956**	**77**	**102**	**143**

TABLE : G07 Velocity, Discharge and Gradients for different Values of Mannings 'n' for Gravity Flow with Pipe Running Full.

Diameter: 400 mm **Area : 0.1257 sqm**

Velocity	*Flow*		*Gradient : 1 : L*		
V, m/s	*ML/d*	*kL/min*	*n = 0.015*	*n = 0.013*	*n = 0.011*
1	*2*	*3*	*4*	*5*	*6*
0.4	4.34	3.01	1289	1717	2398
0.44	4.78	3.32	1066	1419	1982
0.48	5.21	3.62	895	1192	1666
0.52	5.65	3.92	763	1016	1419
0.56	6.08	4.22	658	876	1224
0.6	**6.51**	**4.52**	**573**	**763**	**1066**
0.64	6.95	4.83	504	671	937
0.68	7.38	5.13	446	594	830
0.72	7.82	5.43	398	530	740
0.76	8.25	5.73	357	476	664
0.8	8.69	6.03	322	429	600
0.84	9.12	6.33	292	389	544
0.88	9.55	6.63	266	355	496
0.92	9.99	6.94	244	325	453
0.96	10.42	7.24	224	298	416
1	**10.86**	**7.54**	**206**	**275**	**384**
1.04	11.29	7.84	191	254	355
1.08	11.73	8.15	177	236	329
1.12	12.16	8.44	164	219	306
1.16	12.59	8.74	153	204	285
1.2	13.03	9.05	143	191	266
1.24	13.46	9.35	134	179	250
1.28	13.9	9.65	126	168	234
1.32	14.33	9.95	118	158	220
1.36	14.77	10.26	112	149	207
1.4	15.2	10.56	105	140	196
1.44	15.63	10.85	99	132	185
1.48	16.07	11.16	94	125	175
1.52	16.5	11.46	89	119	166
1.56	16.94	11.76	85	113	158
1.6	**17.37**	**12.06**	**81**	**107**	**150**
1.64	17.81	12.37	77	102	143
1.68	18.24	12.67	73	97	136
1.72	18.68	12.97	70	93	130
1.76	19.11	13.27	67	89	124
1.8	19.54	13.57	64	85	118
1.84	19.98	13.88	61	81	113
1.88	20.41	14.17	58	78	109
1.92	20.85	14.48	56	75	104
1.96	21.28	14.78	54	72	100
2	**21.72**	**15.08**	**52**	**69**	**96**
2.04	22.15	15.38	50	66	92
2.08	22.58	15.68	48	64	89
2.12	23.02	15.99	46	61	85
2.16	23.45	16.28	44	59	82

TABLE: G08 Velocity, Discharge and Gradients for different Values of Mannings 'n' for Gravity Flow with Pipe Running Full.

Diameter: 450 mm **Area: 0.1590 sqm**

Velocity	*Flow*		*Gradient : 1 : L*		
V, m/s	*ML/d*	*kL/min*	*n = 0.015*	*n = 0.013*	*n = 0.011*
1	*2*	*3*	*4*	*5*	*6*
0.4	5.5	3.82	1509	2009	2806
0.44	6.05	4.2	1247	1660	2319
0.48	6.6	4.58	1048	1395	1949
0.52	7.15	4.97	893	1189	1660
0.56	7.7	5.35	770	1025	1432
0.6	**8.24**	**5.72**	**671**	**893**	**1247**
0.64	8.79	6.1	589	785	1096
0.68	9.34	6.49	522	695	971
0.72	9.89	6.87	466	620	866
0.76	10.44	7.25	418	557	777
0.8	10.99	7.63	377	502	702
0.84	11.54	8.01	342	456	636
0.88	12.09	8.4	312	415	580
0.92	12.64	8.78	285	380	530
0.96	13.19	9.16	262	349	487
1	**13.74**	**9.54**	**241**	**321**	**449**
1.04	14.29	9.92	223	297	415
1.08	14.84	10.31	207	276	385
1.12	15.39	10.69	192	256	358
1.16	15.94	11.07	179	239	334
1.2	16.49	11.45	168	223	312
1.24	17.04	11.83	157	209	292
1.28	17.59	12.22	147	196	274
1.32	18.14	12.6	139	184	258
1.36	18.69	12.98	131	174	243
1.4	19.24	13.36	123	164	229
1.44	19.79	13.74	116	155	217
1.48	20.34	14.13	110	147	205
1.52	20.89	14.51	104	139	194
1.56	21.44	14.89	99	132	184
1.6	**21.99**	**15.27**	**94**	**126**	**175**
1.64	22.54	15.65	90	120	167
1.68	23.09	16.03	86	114	159
1.72	23.64	16.42	82	109	152
1.76	24.19	16.8	78	104	145
1.8	24.73	17.17	75	99	139
1.84	25.28	17.56	71	95	133
1.88	25.83	17.94	68	91	127
1.92	26.38	18.32	65	87	122
1.96	26.93	18.7	63	84	117
2	**27.48**	**19.08**	**60**	**80**	**112**
2.04	28.03	19.47	58	77	108
2.08	28.58	19.85	56	74	104
2.12	29.13	20.23	54	72	100
2.16	29.68	20.61	52	69	96

TABLE G09 Velocity, Discharge and Gradients for different Values of Mannings 'n' for Gravity Flow with Pipe Running Full.

Diameter: 500 mm **Area : 0.1964 sqm**

Velocity	*Flow*		*Gradient : 1 : L*		
V, m/s	*ML/d*	*kL/min*	*n = 0.015*	*n = 0.013*	*n = 0.011*
1	2	3	4	5	6
0.4	6.79	4.72	1736	2312	3229
0.44	7.46	5.18	1435	1911	2669
0.48	8.14	5.65	1206	1606	2243
0.52	8.82	6.13	1027	1368	1911
0.56	9.5	6.6	886	1180	1648
0.6	**10.18**	**7.07**	**772**	**1028**	**1435**
0.64	10.86	7.54	678	903	1261
0.68	11.54	8.01	601	800	1117
0.72	12.21	8.48	536	714	997
0.76	12.89	8.95	481	640	895
0.8	13.57	9.42	434	578	807
0.84	14.25	9.9	394	524	732
0.88	14.93	10.37	359	478	667
0.92	15.61	10.84	328	437	610
0.96	16.29	11.31	301	401	561
1	**16.97**	**11.78**	**278**	**370**	**517**
1.04	17.64	12.25	257	342	478
1.08	18.32	12.72	238	317	443
1.12	19	13.19	221	295	412
1.16	19.68	13.67	206	275	384
1.2	20.36	14.14	193	257	359
1.24	21.04	14.61	181	241	336
1.28	21.72	15.08	170	226	315
1.32	22.39	15.55	159	212	297
1.36	23.07	16.02	150	200	279
1.4	23.75	16.49	142	189	264
1.44	24.43	16.97	134	178	249
1.48	25.11	17.44	127	169	236
1.52	25.79	17.91	120	160	224
1.56	26.47	18.38	114	152	212
1.6	**27.14**	**18.85**	**109**	**145**	**202**
1.64	27.82	19.32	103	138	192
1.68	28.5	19.79	98	131	183
1.72	29.18	20.26	94	125	175
1.76	29.86	20.74	90	119	167
1.8	30.54	21.21	86	114	159
1.84	31.22	21.68	82	109	153
1.88	31.89	22.15	79	105	146
1.92	32.57	22.62	75	100	140
1.96	33.25	23.09	72	96	134
2	**33.93**	**23.56**	**69**	**92**	**129**
2.04	34.61	24.03	67	89	124
2.08	35.29	24.51	64	86	119
2.12	35.97	24.98	62	82	115
2.16	36.64	25.44	60	79	111

TABLE : G10 Velocity, Discharge and Gradients for different Values of Mannings '*n*' for Gravity Flow with Pipe Running Full.

Diameter: 550 mm **Area :0.2376 sqm**

Velocity	*Flow*		*Gradient : 1 : L*		
V, m/s	*ML/d*	*kL/min*	*n = 0.015*	*n = 0.013*	*n = 0.011*
1	2	3	4	5	6
0.4	8.21	5.7	1972	2625	3667
0.44	9.03	6.27	1629	2170	3030
0.48	9.85	6.84	1369	1823	2546
0.52	10.67	7.41	1167	1553	2170
0.56	11.5	7.99	1006	1339	1871
0.6	**12.32**	**8.56**	**876**	**1167**	**1630**
0.64	13.14	9.13	770	1025	1432
0.68	13.96	9.69	682	908	1269
0.72	14.78	10.26	608	810	1132
0.76	15.6	10.83	546	727	1016
0.8	16.42	11.4	493	656	917
0.84	17.24	11.97	447	595	831
0.88	18.06	12.54	407	542	758
0.92	18.89	13.12	373	496	693
0.96	19.71	13.69	342	456	637
1	**20.53**	**14.26**	**315**	**420**	**587**
1.04	21.35	14.83	292	388	542
1.08	22.17	15.4	270	360	503
1.12	22.99	15.97	251	335	468
1.16	23.81	16.53	234	312	436
1.2	24.63	17.1	219	292	407
1.24	25.45	17.67	205	273	382
1.28	26.28	18.25	193	256	358
1.32	27.1	18.82	181	241	337
1.36	27.92	19.39	171	227	317
1.4	28.74	19.96	161	214	299
1.44	29.56	20.53	152	203	283
1.48	30.38	21.1	144	192	268
1.52	31.2	21.67	137	182	254
1.56	32.02	22.24	130	173	241
1.6	**32.84**	**22.81**	**123**	**164**	**229**
1.64	33.67	23.38	117	156	218
1.68	34.49	23.95	112	149	208
1.72	35.31	24.52	107	142	198
1.76	36.13	25.09	102	136	189
1.8	36.95	25.66	97	130	181
1.84	37.77	26.23	93	124	173
1.88	38.59	26.8	89	119	166
1.92	39.41	27.37	86	114	159
1.96	40.23	27.94	82	109	153
2	**41.06**	**28.51**	**79**	**105**	**147**
2.04	41.88	29.08	76	101	141
2.08	42.7	29.65	73	97	136
2.12	43.52	30.22	70	93	131
2.16	44.34	30.79	68	90	126

TABLE : G11 Velocity, Discharge and Gradients for different Values of Mannings 'n' for Gravity Flow with Pipe Running Full.

Diameter: 600 mm **Area : 0.2827 sqm**

Velocity	Flow		Gradient : 1 : L		
V, m/s	ML/d	kL/min	n = 0.015	n = 0.013	n = 0.011
1	2	3	4	5	6
0.4	**9.77**	**6.78**	**2214**	**2948**	**4118**
0.44	10.75	7.47	1830	2436	3403
0.48	11.73	8.15	1538	2047	2860
0.52	12.7	8.82	1310	1744	2436
0.56	13.68	9.5	1130	1504	2101
0.6	**14.66**	**10.18**	**984**	**1310**	**1830**
0.64	15.63	10.85	865	1152	1608
0.68	16.61	11.53	766	1020	1425
0.72	17.59	12.22	683	910	1271
0.76	18.57	12.9	613	817	1141
0.8	19.54	13.57	554	737	1029
0.84	20.52	14.25	502	668	934
0.88	21.5	14.93	457	609	851
0.92	22.48	15.61	419	557	778
0.96	23.45	16.28	384	512	715
1	**24.43**	**16.97**	**354**	**472**	**659**
1.04	25.41	17.65	328	436	609
1.08	26.38	18.32	304	404	565
1.12	27.36	19	282	376	525
1.16	28.34	19.68	263	351	490
1.2	29.32	20.36	246	328	458
1.24	30.29	21.03	230	307	428
1.28	31.27	21.72	216	288	402
1.32	32.25	22.4	203	271	378
1.36	33.22	23.07	192	255	356
1.4	34.2	23.75	181	241	336
1.44	35.18	24.43	171	227	318
1.48	36.16	25.11	162	215	301
1.52	37.13	25.78	153	204	285
1.56	38.11	26.47	146	194	271
1.6	**39.09**	**27.15**	**138**	**184**	**257**
1.64	40.06	27.82	132	175	245
1.68	41.04	28.5	126	167	233
1.72	42.02	29.18	120	159	223
1.76	43	29.86	114	152	213
1.8	43.97	30.53	109	146	203
1.84	44.95	31.22	105	139	195
1.88	45.93	31.9	100	133	186
1.92	46.9	32.57	96	128	179
1.96	47.88	33.25	92	123	171
2	**48.86**	**33.93**	**89**	**118**	**165**
2.04	49.84	34.61	85	113	158
2.08	50.81	35.28	82	109	152
2.12	51.79	35.97	79	105	147
2.16	52.77	36.65	76	101	141

TABLE : G12 Velocity, Discharge and Gradients for different Values of Mannings 'n' for Gravity Flow with Pipe Running Full.

Diameter: 650 mm **Area : 0.3318 sqm**

Velocity	*Flow*		*Gradient : 1 : L*		
V, m/s	*ML/d*	*kL/min*	*n = 0.015*	*n = 0.013*	*n = 0.011*
1	2	3	4	5	6
0.4	11.47	7.97	2463	3280	4581
0.44	12.62	8.76	2036	2711	3786
0.48	13.76	9.56	1711	2278	3181
0.52	14.91	10.35	1458	1941	2711
0.56	16.06	11.15	1257	1673	2337
0.6	**17.2**	**11.94**	**1095**	**1458**	**2036**
0.64	18.35	12.74	962	1281	1790
0.68	19.5	13.54	852	1135	1585
0.72	20.64	14.33	760	1012	1414
0.76	21.79	15.13	682	909	1269
0.8	22.94	15.93	616	820	1145
0.84	24.08	16.72	559	744	1039
0.88	25.23	17.52	509	678	947
0.92	26.38	18.32	466	620	866
0.96	27.52	19.11	428	569	795
1	**28.67**	**19.91**	**394**	**525**	**733**
1.04	29.82	20.71	364	485	678
1.08	30.96	21.5	338	450	628
1.12	32.11	22.3	314	418	584
1.16	33.26	23.1	293	390	545
1.2	34.41	23.9	274	364	509
1.24	35.55	24.69	256	341	477
1.28	36.7	25.49	241	320	447
1.32	37:85	26.28	226	301	421
1.36	38.99	27.08	213	284	396
1.4	40.14	27.88	201	268	374
1.44	41.29	28.67	190	253	353
1.48	42.43	29.47	180	240	335
1.52	43.58	30.26	171	227	317
1.56	44.73	31.06	162	216	301
1.6	**45.87**	**31.85**	**154**	**205**	**286**
1.64	47.02	32.65	147	195	273
1.68	48.17	33.45	140	186	260
1.72	49.31	34.24	133	177	248
1.76	50.46	35.04	127	169	237
1.8	51.61	35.84	122	162	226
1.84	52.75	36.63	116	155	217
1.88	53.9	37.43	112	148	207
1.92	55.05	38.23	107	142	199
1.96	56.19	39.02	103	137	191
2	**57.34**	**39.82**	**99**	**131**	**183**
2.04	58.49	40.62	95	126	176
2.08	59.64	41.42	91	121	169
2.12	60.78	42.21	88	117	163
2.16	61.93	43.01	84	112	157

TABLE : G13 Velocity, Discharge and Gradients for different Values of Mannings 'n' for Gravity Flow with Pipe Running Full.

Diameter: 700 mm **Area : 0.3848 sqm**

Velocity	*Flow*		*Gradient : 1 : L*		
V, m/s	*ML/d*	*kL/min*	*n = 0.015*	*n = 0.013*	*n = 0.011*
1	*2*	*3*	*4*	*5*	*6*
0.4	13.3	9.24	2719	3621	5057
0.44	14.63	10.16	2247	2992	4179
0.48	15.96	11.08	1888	2514	3512
0.52	17.29	12.01	1609	2142	2992
0.56	18.62	12.93	1387	1847	2580
0.6	**19.95**	**13.85**	**1209**	**1609**	**2248**
0.64	21.28	14.78	1062	1414	1975
0.68	22.61	15.7	941	1253	1750
0.72	23.94	16.63	839	1117	1561
0.76	25.27	17.55	753	1003	1401
0.8	26.6	18.47	680	905	1264
0.84	27.93	19.4	617	821	1147
0.88	29.26	20.32	562	748	1045
0.92	30.59	21.24	514	684	956
0.96	31.92	22.17	472	629	878
1	**33.25**	**23.09**	**435**	**579**	**809**
1.04	34.58	24.01	402	536	748
1.08	35.91	24.94	373	497	694
1.12	37.24	25.86	347	462	645
1.16	38.57	26.78	323	431	601
1.2	39.9	27.71	302	402	562
1.24	41.23	28.63	283	377	526
1.28	42.56	29.56	266	354	494
1.32	43.89	30.48	250	332	464
1.36	45.22	31.4	235	313	437
1.4	46.55	32.33	222	296	413
1.44	47.88	33.25	210	279	390
1.48	49.21	34.17	199	264	369
1.52	50.54	35.1	188	251	350
1.56	51.87	36.02	179	238	332
1.6	**53.2**	**36.94**	**170**	**226**	**316**
1.64	54.53	37.87	162	215	301
1.68	55.86	38.79	154	205	287
1.72	57.19	39.72	147	196	273
1.76	58.52	40.64	140	187	261
1.8	59.85	41.56	134	179	250
1.84	61.18	42.49	129	171	239
1.88	62.51	43.41	123	164	229
1.92	63.84	44.33	118	157	219
1.96	65.17	45.26	113	151	211
2	**66.5**	**46.18**	**109**	**145**	**202**
2.04	67.83	47.1	105	139	194
2.08	69.16	48.03	101	134	187
2.12	70.49	48.95	97	129	180
2.16	71.82	49.88	93	124	173

TABLE : G14 Velocity, Discharge and Gradients for different Values of Mannings 'n' for Gravity Flow with Pipe Running Full.

Diameter: 750 mm **Area : 0.4418 sqm**

Velocity	*Flow*		*Gradient : 1 : L*		
V, m/s	*ML/d*	*kL/min*	*n = 0.015*	*n = 0.013*	*n = 0.011*
1	*2*	*3*	*4*	*5*	*6*
0.4	15.27	10.6	2981	3969	5544
0.44	16.8	11.67	2464	3280	4582
0.48	18.32	12.72	2070	2756	3850
0.52	19.85	13.78	1764	2349	3281
0.56	21.38	14.85	1521	2025	2829
0.6	**22.9**	**15.9**	**1325**	**1764**	**2464**
0.64	24.43	16.97	1165	1551	2166
0.68	25.96	18.03	1032	1373	1918
0.72	27.48	19.08	920	1225	1711
0.76	29.01	20.15	826	1100	1536
0.8	30.54	21.21	745	992	1386
0.84	32.06	22.26	676	900	1257
0.88	33.59	23.33	616	820	1145
0.92	35.12	24.39	564	750	1048
0.96	36.64	25.44	518	689	963
1	**38.17**	**26.51**	**477**	**635**	**887**
1.04	39.7	27.57	441	587	820
1.08	41.22	28.63	409	544	761
1.12	42.75	29.69	380	506	707
1.16	44.28	30.75	354	472	659
1.2	45.81	31.81	331	441	616
1.24	47.33	32.87	310	413	577
1.28	48.86	33.93	291	388	541
1.32	50.39	34.99	274	364	509
1.36	51.91	36.05	258	343	480
1.4	53.44	37.11	243	324	453
1.44	54.97	38.17	230	306	428
1.48	56.49	39.23	218	290	405
1.52	58.02	40.29	206	275	384
1.56	59.55	41.35	196	261	365
1.6	**61.07**	**42.41**	**186**	**248**	**347**
1.64	62.6	43.47	177	236	330
1.68	64.13	44.53	169	225	314
1.72	65.65	45.59	161	215	300
1.76	67.18	46.65	154	205	286
1.8	68.71	47.72	147	196	274
1.84	70.24	48.78	141	188	262
1.88	71.76	49.83	135	180	251
1.92	73.29	50.9	129	172	241
1.96	74.82	51.96	124	165	231
2	**76.34**	**53.01**	**119**	**159**	**222**
2.04	77.87	54.08	115	153	213
2.08	79.4	55.14	110	147	205
2.12	80.92	56.19	106	141	197
2.16	82.45	57.26	102	136	190

TABLE : G15 Velocity, Discharge and Gradients for different Values of Mannings 'n' for Gravity Flow with Pipe Running Full.

Diameter: 800 mm **Area : 0.5027 sqm**

Velocity	*Flow*		*Gradient : 1 : L*		
V, m/s	*ML/d*	*kL/min*	*n = 0.015*	*n = 0.013*	*n = 0.011*
1	2	3	4	5	6
0.4	17.37	12.06	3249	4326	6042
0.44	19.11	13.27	2685	3575	4994
0.48	20.85	14.48	2256	3004	4196
0.52	22.58	15.68	1923	2560	3575
0.56	24.32	16.89	1658	2207	3083
0.6	**26.06**	**18.1**	**1444**	**1923**	**2685**
0.64	27.8	19.31	1269	1690	2360
0.68	29.53	20.51	1124	1497	2091
0.72	31.27	21.72	1003	1335	1865
0.76	33.01	22.92	900	1198	1674
0.8	34.74	24.13	812	1081	1511
0.84	36.48	25.33	737	981	1370
0.88	38.22	26.54	671	894	1248
0.92	39.96	27.75	614	818	1142
0.96	41.69	28.95	564	751	1049
1	**43.43**	**30.16**	**520**	**692**	**967**
1.04	45.17	31.37	481	640	894
1.08	46.9	32.57	446	593	829
1.12	48.64	33.78	414	552	771
1.16	50.38	34.99	386	514	718
1.2	52.12	36.19	361	481	671
1.24	53.85	37.4	338	450	629
1.28	55.59	38.6	317	422	590
1.32	57.33	39.81	298	397	555
1.36	59.07	41.02	281	374	523
1.4	60.8	42.22	265	353	493
1.44	62.54	43.43	251	334	466
1.48	64.28	44.64	237	316	441
1.52	66.01	45.84	225	300	418
1.56	67.75	47.05	214	284	397
1.6	**69.49**	**48.26**	**203**	**270**	**378**
1.64	71.23	49.47	193	257	359
1.68	72.96	50.67	184	245	343
1.72	74.7	51.88	176	234	327
1.76	76.44	53.08	168	223	312
1.8	78.17	54.28	160	214	298
1.84	79.91	55.49	154	204	286
1.88	81.65	56.7	147	196	274
1.92	83.39	57.91	141	188	262
1.96	85.12	59.11	135	180	252
2	**86.86**	**60.32**	**130**	**173**	**242**
2.04	88.6	61.53	125	166	232
2.08	90.34	62.74	120	160	223
2.12	92.07	63.94	116	154	215
2.16	93.81	65.15	111	148	207

TABLE : G16 Velocity, Discharge and Gradients for different Values of Mannings 'n' for Gravity Flow with Pipe Running Full.

Diameter: 850 mm **Area : 0.5674 sqm**

Velocity	*Flow*		*Gradient : 1 : L*		
V, m/s	*ML/d*	*kL/min*	*n = 0.015*	*n = 0.013*	*n = 0.011*
1	2	3	4	5	6
0.4	19.61	13.62	3523	4690	6551
0.44	21.57	14.98	2911	3876	5414
0.48	23.53	16.34	2446	3257	4549
0.52	25.5	17.71	2084	2775	3876
0.56	27.46	19.07	1797	2393	3342
0.6	**29.42**	**20.43**	**1566**	**2084**	**2911**
0.64	31.38	21.79	1376	1832	2559
0.68	33.34	23.15	1219	1623	2267
0.72	35.3	24.51	1087	1448	2022
0.76	37.26	25.88	976	1299	1815
0.8	39.22	27.24	881	1173	1638
0.84	41.18	28.6	799	1064	1485
0.88	43.15	29.97	728	969	1353
0.92	45.11	31.33	666	887	1238
0.96	47.07	32.69	612	814	1137
1	**49.03**	**34.05**	**564**	**750**	**1048**
1.04	50.99	35.41	521	694	969
1.08	52.95	36.77	483	643	899
1.12	54.91	38.13	449	598	836
1.16	56.87	39.49	419	558	779
1.2	58.83	40.85	391	521	728
1.24	60.8	42.22	367	488	682
1.28	62.76	43.58	344	458	640
1.32	64.72	44.94	323	431	602
1.36	66.68	46.31	305	406	567
1.4	68.64	47.67	288	383	535
1.44	70.6	49.03	272	362	505
1.48	72.56	50.39	257	343	479
1.52	74.52	51.75	244	325	454
1.56	76.49	53.12	232	308	431
1.6	**78.45**	**54.48**	**220**	**293**	**409**
1.64	80.41	55.84	210	279	390
1.68	82.37	57.2	200	266	371
1.72	84.33	58.56	191	254	354
1.76	86.29	59.92	182	242	338
1.8	88.25	61.28	174	232	323
1.84	90.21	62.65	166	222	310
1.88	92.17	64.01	159	212	297
1.92	94.14	65.38	153	204	284
1.96	96.1	66.74	147	195	273
2	**98.06**	**68.1**	**141**	**188**	**262**
2.04	100.02	69.46	135	180	252
2.08	101.98	70.82	130	173	242
2.12	103.94	72.18	125	167	233
2.16	105.9	73.54	121	161	225

TABLE : G17 Velocity, Discharge and Gradients for different Values of Mannings 'n' for Gravity Flow with Pipe Running Full.

Diameter: 900 mm **Area : 0.6362 sqm**

Velocity	*Flow*		*Gradient : 1 : L*		
V, m/s	*ML/d*	*kL/min*	*n = 0.015*	*n = 0.013*	*n = 0.011*
1	*2*	*3*	*4*	*5*	*6*
0.4	21.99	15.27	3802	5061	7069
0.44	24.19	16.8	3142	4183	5842
0.48	26.38	18.32	2640	3515	4909
0.52	28.58	19.85	2249	2995	4183
0.56	30.78	21.38	1940	2582	3607
0.6	**32.98**	**22.9**	**1690**	**2249**	**3142**
0.64	35.18	24.43	1485	1977	2761
0.68	37.38	25.96	1315	1751	2446
0.72	39.58	27.49	1173	1562	2182
0.76	41.77	29.01	1053	1402	1958
0.8	43.97	30.53	950	1265	1767
0.84	46.17	32.06	862	1148	1603
0.88	48.37	33.59	785	1046	1461
0.92	50.57	35.12	719	957	1336
0.96	52.77	36.65	660	879	1227
1	**54.97**	**38.17**	**608**	**810**	**1131**
1.04	57.17	39.7	562	749	1046
1.08	59.36	41.22	521	694	970
1.12	61.56	42.75	485	646	902
1.16	63.76	44.28	452	602	841
1.2	65.96	45.81	422	562	785
1.24	68.16	47.33	396	527	736
1.28	70.36	48.86	371	494	690
1.32	72.56	50.39	349	465	649
1.36	74.75	51.91	329	438	612
1.4	76.95	53.44	310	413	577
1.44	79.15	54.97	293	391	545
1.48	81.35	56.49	278	370	516
1.52	83.55	58.02	263	351	490
1.56	85.75	59.55	250	333	465
1.6	**87.95**	**61.08**	**238**	**316**	**442**
1.64	90.15	62.6	226	301	421
1.68	92.34	64.13	216	287	401
1.72	94.54	65.65	206	274	382
1.76	96.74	67.18	196	261	365
1.8	98.94	68.71	188	250	349
1.84	101.14	70.24	180	239	334
1.88	103.34	71.76	172	229	320
1.92	105.54	73.29	165	220	307
1.96	107.73	74.81	158	211	294
2	**109.93**	**76.34**	**152**	**202**	**283**
2.04	112.13	77.87	146	195	272
2.08	114.33	79.4	141	187	261
2.12	116.53	80.92	135	180	252
2.16	118.73	82.45	130	174	242

TABLE : G18 Velocity, Discharge and Gradients for different Values of Mannings 'n' for Gravity Flow with Pipe Running Full.

Diameter: 950 mm **Area : 0.7088 sqm**

Velocity V, m/s	*Flow*		*Gradient : 1 : L*		
	ML/d	*kL/min*	*n = 0.015*	*n = 0.013*	*n = 0.011*
1	*2*	*3*	*4*	*5*	*6*
0.4	24.5	17.01	4086	5440	7598
0.44	26.95	18.72	3377	4496	6279
0.48	29.4	20.42	2837	3778	5276
0.52	31.85	22.12	2418	3219	4496
0.56	34.3	23.82	2085	2775	3876
0.6	**36.75**	**25.52**	**1816**	**2418**	**3377**
0.64	39.2	27.22	1596	2125	2968
0.68	41.65	28.92	1414	1882	2629
0.72	44.1	30.63	1261	1679	2345
0.76	46.55	32.33	1132	1507	2105
0.8	48.99	34.02	1021	1360	1899
0.84	51.44	35.72	926	1233	1723
0.88	53.89	37.42	844	1124	1570
0.92	56.34	39.13	772	1028	1436
0.96	58.79	40.83	709	944	1319
1	**61.24**	**42.53**	**654**	**870**	**1216**
1.04	63.69	44.23	604	805	1124
1.08	66.14	45.93	560	746	1042
1.12	68.59	47.63	521	694	969
1.16	71.04	49.33	486	647	903
1.2	73.49	51.03	454	604	844
1.24	75.94	52.74	425	566	791
1.28	78.39	54.44	399	531	742
1.32	80.84	56.14	375	500	698
1.36	83.29	57.84	353	471	657
1.4	85.74	59.54	334	444	620
1.44	88.19	61.24	315	420	586
1.48	90.64	62.94	298	397	555
1.52	93.09	64.65	283	377	526
1.56	95.54	66.35	269	358	500
1.6	**97.99**	**68.05**	**255**	**340**	**475**
1.64	100.44	69.75	243	324	452
1.68	102.89	71.45	232	308	431
1.72	105.34	73.15	221	294	411
1.76	107.79	74.85	211	281	392
1.8	110.24	76.56	202	269	375
1.84	112.69	78.26	193	257	359
1.88	115.14	79.96	185	246	344
1.92	117.59	81.66	177	236	330
1.96	120.04	83.36	170	227	316
2	**122.49**	**85.06**	**163**	**218**	**304**
2.04	124.94	86.76	157	209	292
2.08	127.39	88.47	151	201	281
2.12	129.84	90.17	145	194	270
2.16	132.29	91.87	140	187	261

TABLE : G19 Velocity, Discharge and Gradients for different Values of Mannings 'n' for Gravity Flow with Pipe Running Full.

Diameter: 1000 mm **Area : 0.7854 sqm**

Velocity	*Flow*		*Gradient : 1 : L*		
V, m/s	*ML/d*	*kL/min*	*n = 0.015*	*n = 0.013*	*n = 0.011*
1	*2*	*3*	*4*	*5*	*6*
0.4	27.14	18.85	4375	5825	8135
0.44	29.86	20.74	3616	4814	6723
0.48	32.57	22.62	3038	4045	5650
0.52	35.29	24.51	2589	3446	4814
0.56	38	26.39	2232	2972	4151
0.6	**40.72**	**28.28**	**1944**	**2589**	**3616**
0.64	43.43	30.16	1709	2275	3178
0.68	46.14	32.04	1514	2015	2815
0.72	48.86	33.93	1350	1798	2511
0.76	51.57	35.81	1212	1613	2254
0.8	54.29	37.7	1094	1456	2034
0.84	57	39.58	992	1321	1845
0.88	59.72	41.47	904	1203	1681
0.92	62.43	43.35	827	1101	1538
0.96	65.15	45.24	760	1011	1412
1	**67.86**	**47.13**	**700**	**932**	**1302**
1.04	70.57	49.01	647	862	1203
1.08	73.29	50.9	600	799	1116
1.12	76	52.78	558	743	1038
1.16	78.72	54.67	520	693	967
1.2	81.43	56.55	486	647	904
1.24	84.15	58.44	455	606	847
1.28	86.86	60.32	427	569	794
1.32	89.58	62.21	402	535	747
1.36	92.29	64.09	378	504	704
1.4	95	65.97	357	475	664
1.44	97.72	67.86	338	449	628
1.48	100.43	69.74	320	425	594
1.52	103.15	71.63	303	403	563
1.56	105.86	73.51	288	383	535
1.6	**108.58**	**75.4**	**273**	**364**	**508**
1.64	111.29	77.28	260	346	484
1.68	114	79.17	248	330	461
1.72	116.72	81.06	237	315	440
1.76	119.43	82.94	226	301	420
1.8	122.15	84.83	216	288	402
1.84	124.86	86.71	207	275	384
1.88	127.58	88.6	198	264	368
1.92	130.29	90.48	190	253	353
1.96	133.01	92.37	182	243	339
2	**135.72**	**94.25**	**175**	**233**	**325**
2.04	138.43	96.13	168	224	313
2.08	141.15	98.02	162	215	301
2.12	143.86	99.9	156	207	290
2.16	146.58	101.79	150	200	279

TABLE : G20 Velocity, Discharge and Gradients for different Values of Mannings 'n' for Gravity Flow with Pipe Running Full.

Diameter: 1100 mm **Area : 0.9503 sqm**

Velocity	*Flow*		*Gradient : 1 : L*		
V, m/s	*ML/d*	*kL/min*	*n = 0.015*	*n = 0.013*	*n = 0.011*
1	*2*	*3*	*4*	*5*	*6*
0.4	32.84	22.81	4968	6614	9237
0.44	36.13	25.09	4106	5466	7634
0.48	39.41	27.37	3450	4593	6415
0.52	42.7	29.65	2940	3913	5466
0.56	45.98	31.93	2535	3374	4713
0.6	**49.27**	**34.22**	**2208**	**2939**	**4106**
0.64	52.55	36.49	1941	2583	3608
0.68	55.84	38.78	1719	2288	3196
0.72	59.12	41.06	1533	2041	2851
0.76	62.4	43.33	1376	1832	2559
0.8	65.69	45.62	1242	1653	2309
0.84	68.97	47.9	1126	1500	2095
0.88	72.26	50.18	1026	1366	1909
0.92	75.54	52.46	939	1250	1746
0.96	78.83	54.74	862	1148	1604
1	**82.11**	**57.02**	**795**	**1058**	**1478**
1.04	85.4	59.31	735	978	1366
1.08	88.68	61.58	681	907	1267
1.12	91.96	63.86	634	844	1178
1.16	95.25	66.15	591	786	1098
1.2	98.53	68.42	552	735	1026
1.24	101.82	70.71	517	688	961
1.28	105.1	72.99	485	646	902
1.32	108.39	75.27	456	607	848
1.36	111.67	77.55	430	572	799
1.4	114.95	79.83	406	540	754
1.44	118.24	82.11	383	510	713
1.48	121.52	84.39	363	483	675
1.52	124.81	86.67	344	458	640
1.56	128.09	88.95	327	435	607
1.6	**131.38**	**91.24**	**310**	**413**	**577**
1.64	134.66	93.51	296	393	550
1.68	137.95	95.8	282	375	524
1.72	141.23	98.08	269	358	500
1.76	144.51	100.35	257	342	477
1.8	147.8	102.64	245	327	456
1.84	151.08	104.92	235	313	437
1.88	154.37	107.2	225	299	418
1.92	157.65	109.48	216	287	401
1.96	160.94	111.76	207	275	385
2	**164.22**	**114.04**	**199**	**265**	**369**
2.04	167.51	116.33	191	254	355
2.08	170.79	118.6	184	245	342
2.12	174.07	120.88	177	235	329
2.16	177.36	123.17	170	227	317

TABLE : G21 Velocity, Discharge and Gradients for different Values of Mannings 'n' for Gravity Flow with Pipe Running Full.

Diameter: 1200 mm **Area :1.1310 sqm**

Velocity V, m/s	Flow ML/d	Flow kL/min	Gradient : 1 : L n = 0.015	Gradient : 1 : L n = 0.013	Gradient : 1 : L n = 0.011
1	2	3	4	5	6
0.4	39.09	27.15	5579	7427	10374
0.44	43	29.86	4611	6138	8573
0.48	46.9	32.57	3874	5158	7204
0.52	50.81	35.28	3301	4395	6138
0.56	54.72	38	2846	3789	5293
0.6	**58.63**	**40.72**	**2479**	**3301**	**4610**
0.64	62.54	43.43	2179	2901	4052
0.68	66.45	46.15	1930	2570	3589
0.72	70.36	48.86	1722	2292	3202
0.76	74.27	51.58	1545	2057	2874
0.8	78.17	54.28	1395	1857	2593
0.84	82.08	57	1265	1684	2352
0.88	85.99	59.72	1153	1534	2143
0.92	89.9	62.43	1055	1404	1961
0.96	93.81	65.15	969	1289	1801
1	**97.72**	**67.86**	**893**	**1188**	**1660**
1.04	101.63	70.58	825	1099	1535
1.08	105.54	73.29	765	1019	1423
1.12	109.44	76	712	947	1323
1.16	113.35	78.72	663	883	1233
1.2	117.26	81.43	620	825	1153
1.24	121.17	84.15	581	773	1079
1.28	125.08	86.86	545	725	1013
1.32	128.99	89.58	512	682	953
1.36	132.9	92.29	483	642	897
1.4	136.81	95.01	455	606	847
1.44	140.71	97.72	430	573	800
1.48	144.62	100.43	408	543	758
1.52	148.53	103.15	386	514	718
1.56	152.44	105.86	367	488	682
1.6	**156.35**	**108.58**	**349**	**464**	**648**
1.64	160.26	111.29	332	442	617
1.68	164.17	114.01	316	421	588
1.72	168.08	116.72	302	402	561
1.76	171.98	119.43	288	384	536
1.8	175.89	122.15	275	367	512
1.84	179.8	124.86	264	351	490
1.88	183.71	127.58	253	336	470
1.92	187.62	130.29	242	322	450
1.96	191.53	133.01	232	309	432
2	**195.44**	**135.72**	**223**	**297**	**415**
2.04	199.35	138.44	214	286	399
2.08	203.25	141.15	206	275	384
2.12	207.16	143.86	199	264	369
2.16	211.07	146.58	191	255	356

TABLE : G22 Velocity, Discharge and Gradients for different Values of Mannings 'n' for Gravity Flow with Pipe Running Full.

Diameter: 1300 mm **Area : 1.3273 sqm**

Velocity	*Flow*		*Gradient : 1 : L*		
V, m/s	*ML/d*	*kL/min*	*n = 0.015*	*n = 0.013*	*n = 0.011*
1	2	3	4	5	6
0.4	**45.87**	**31.85**	**6207**	**8263**	**11542**
0.43	49.31	34.24	5371	7150	9987
0.46	52.75	36.63	4694	6248	8727
0.49	56.19	39.02	4136	5507	7691
0.52	59.64	41.42	3673	4889	6829
0.55	63.08	43.81	3283	4371	6105
0.58	66.52	46.19	2952	3930	5489
0.61	69.96	48.58	2669	3553	4963
0.64	73.4	50.97	2425	3228	4508
0.67	76.84	53.36	2212	2945	4114
0.7	**80.28**	**55.75**	**2027**	**2698**	**3769**
0.73	83.72	58.14	1864	2481	3465
0.76	87.16	60.53	1719	2289	3197
0.79	90.6	62.92	1591	2118	2959
0.82	94.04	65.31	1477	1966	2746
0.85	97.48	67.69	1375	1830	2556
0.88	100.92	70.08	1282	1707	2385
0.91	104.36	72.47	1199	1597	2230
0.94	107.8	74.86	1124	1496	2090
0.97	111.24	77.25	1056	1405	1963
1	**114.68**	**79.64**	**993**	**1322**	**1847**
1.03	118.12	82.03	936	1246	1741
1.06	121.56	84.42	884	1177	1644
1.09	125	86.81	836	1113	1554
1.12	128.45	89.2	792	1054	1472
1.15	131.89	91.59	751	1000	1396
1.18	135.33	93.98	713	950	1326
1.21	138.77	96.37	678	903	1261
1.24	142.21	98.76	646	860	1201
1.27	145.65	101.15	616	820	1145
1.3	**149.09**	**103.53**	**588**	**782**	**1093**
1.33	152.53	105.92	561	747	1044
1.36	155.97	108.31	537	715	998
1.39	159.41	110.7	514	684	956
1.42	162.85	113.09	493	656	916
1.45	166.29	115.48	472	629	878
1.48	169.73	117.87	453	604	843
1.51	173.17	120.26	436	580	810
1.54	176.61	122.65	419	557	779
1.57	180.05	125.03	403	536	749
1.6	**183.49**	**127.42**	**388**	**516**	**721**
1.63	186.93	129.81	374	498	695
1.66	190.37	132.2	360	480	670
1.69	193.81	134.59	348	463	647
1.72	197.26	136.99	336	447	624

TABLE : G23 Velocity, Discharge and Gradients for different Values of Mannings 'n' for Gravity Flow with Pipe Running Full.

Diameter: 1400 mm **Area : 1.5394 sqm**

Velocity	*Flow*		*Gradient : 1 : L*		
V, m/s	*ML/d*	*kL/min*	*n = 0.015*	*n = 0.013*	*n = 0.011*
1	2	3	4	5	6
0.4	**53.2**	**36.94**	**6852**	**9121**	**12740**
0.43	57.19	39.72	5929	7893	11024
0.46	61.18	42.49	5181	6897	9633
0.49	65.17	45.26	4566	6078	8490
0.52	69.16	48.03	4054	5397	7538
0.55	73.15	50.8	3624	4824	6738
0.58	77.14	53.57	3259	4338	6059
0.61	81.13	56.34	2946	3922	5478
0.64	85.12	59.11	2676	3563	4977
0.67	89.11	61.88	2442	3251	4541
0.7	**93.1**	**64.65**	**2237**	**2978**	**4160**
0.73	97.09	67.42	2057	2739	3825
0.76	101.08	70.19	1898	2527	3529
0.79	105.07	72.97	1757	2338	3266
0.82	109.06	75.74	1630	2170	3032
0.85	113.05	78.51	1517	2020	2821
0.88	117.04	81.28	1416	1885	2632
0.91	121.04	84.06	1324	1762	2462
0.94	125.03	86.83	1241	1652	2307
0.97	129.02	89.6	1165	1551	2166
1	**133.01**	**92.37**	**1096**	**1459**	**2038**
1.03	137	95.14	1033	1376	1921
1.06	140.99	97.91	976	1299	1814
1.09	144.98	100.68	923	1228	1716
1.12	148.97	103.45	874	1163	1625
1.15	152.96	106.22	829	1104	1541
1.18	156.95	108.99	787	1048	1464
1.21	160.94	111.76	749	997	1392
1.24	164.93	114.53	713	949	1326
1.27	168.92	117.31	680	905	1264
1.3	**172.91**	**120.08**	**649**	**864**	**1206**
1.33	176.9	122.85	620	825	1152
1.36	180.89	125.62	593	789	1102
1.39	184.88	128.39	567	755	1055
1.42	188.87	131.16	544	724	1011
1.45	192.86	133.93	521	694	970
1.48	196.85	136.7	500	666	931
1.51	200.84	139.47	481	640	894
1.54	204.83	142.24	462	615	859
1.57	208.82	145.01	445	592	827
1.6	**212.81**	**147.78**	**428**	**570**	**796**
1.63	216.8	150.56	413	549	767
1.66	220.79	153.33	398	530	740
1.69	224.78	156.1	384	511	714
1.72	228.77	158.87	371	493	689

TABLE : G24 Velocity, Discharge and Gradients for different Values of Mannings 'n' for Gravity Flow with Pipe Running Full.

Diameter: 1500 mm **Area : 1.7672 sqm**

Velocity	Flow		Gradient : 1 : L		
V, m/s	*ML/d*	*kL/min*	*n = 0.015*	*n = 0.013*	*n = 0.011*
1	2	3	4	5	6
0.4	**61.07**	**42.41**	**7512**	**10000**	**13967**
0.43	65.65	45.59	6500	8653	12086
0.46	70.24	48.78	5680	7561	10561
0.49	74.82	51.96	5006	6664	9308
0.52	79.4	55.14	4445	5917	8265
0.55	83.98	58.32	3973	5289	7388
0.58	88.56	61.5	3573	4756	6643
0.61	93.14	64.68	3230	4300	6006
0.64	97.72	67.86	2934	3906	5456
0.67	102.3	71.04	2677	3564	4978
0.7	**106.88**	**74.22**	**2453**	**3265**	**4561**
0.73	111.46	77.4	2255	3002	4194
0.76	116.04	80.58	2081	2770	3869
0.79	120.62	83.76	1926	2564	3581
0.82	125.2	86.94	1788	2379	3324
0.85	129.78	90.13	1664	2214	3093
0.88	134.36	93.31	1552	2066	2886
0.91	138.94	96.49	1451	1932	2699
0.94	143.52	99.67	1360	1811	2529
0.97	148.1	102.85	1277	1700	2375
1	**152.69**	**106.03**	**1202**	**1600**	**2235**
1.03	157.27	109.22	1133	1508	2106
1.06	161.85	112.4	1070	1424	1989
1.09	166.43	115.58	1012	1347	1881
1.12	171.01	118.76	958	1275	1782
1.15	175.59	121.94	909	1210	1690
1.18	180.17	125.12	863	1149	1605
1.21	184.75	128.3	821	1093	1526
1.24	189.33	131.48	782	1041	1453
1.27	193.91	134.66	745	992	1386
1.3	**198.49**	**137.84**	**711**	**947**	**1322**
1.33	203.07	141.02	679	904	1263
1.36	207.65	144.2	650	865	1208
1.39	212.23	147.38	622	828	1157
1.42	216.81	150.56	596	793	1108
1.45	221.39	153.74	572	761	1063
1.48	225.97	156.92	549	730	1020
1.51	230.55	160.1	527	702	980
1.54	235.13	163.28	507	675	942
1.57	239.72	166.47	488	649	907
1.6	**244.3**	**169.65**	**470**	**625**	**873**
1.63	248.88	172.83	452	602	841
1.66	253.46	176.01	436	581	811
1.69	258.04	179.19	421	560	782
1.72	262.62	182.38	406	541	755

TABLE : G25 Velocity, Discharge and Gradients for different Values of Mannings 'n' for Gravity Flow with Pipe Running Full.

Diameter: 1600 mm **Area : 2.0106 sqm**

Velocity	*Flow*		*Gradient : 1 : L*		
V, m/s	*ML/d*	*kL/min*	*n = 0.015*	*n = 0.013*	*n = 0.011*
1	*2*	*3*	*4*	*5*	*6*
0.4	69.49	48.26	8187	10898	15222
0.425	73.83	51.27	7252	9654	13484
0.45	78.17	54.28	6469	8611	12027
0.475	82.52	57.31	5806	7728	10794
0.5	**86.86**	**60.32**	**5240**	**6975**	**9742**
0.525	91.2	63.33	4753	6326	8836
0.55	95.55	66.35	4330	5764	8051
0.575	99.89	69.37	3962	5274	7366
0.6	104.23	72.38	3639	4844	6765
0.625	108.58	75.4	3353	4464	6235
0.65	112.92	78.42	3100	4127	5765
0.675	117.26	81.43	2875	3827	5345
0.7	121.61	84.45	2673	3559	4970
0.725	125.95	87.47	2492	3317	4634
0.75	**130.29**	**90.48**	**2329**	**3100**	**4330**
0.775	134.63	93.49	2181	2903	4055
0.8	138.98	96.51	2047	2725	3805
0.825	143.32	99.53	1925	2562	3578
0.85	147.66	102.54	1813	2413	3371
0.875	152.01	105.56	1711	2277	3181
0.9	156.35	108.58	1617	2153	3007
0.925	160.69	111.59	1531	2038	2846
0.95	165.04	114.61	1451	1932	2699
0.975	169.38	117.63	1378	1834	2562
1	**173.72**	**120.64**	**1310**	**1744**	**2436**
1.025	178.06	123.65	1247	1660	2318
1.05	182.41	126.67	1188	1582	2209
1.075	186.75	129.69	1134	1509	2108
1.1	191.09	132.7	1083	1441	2013
1.125	195.44	135.72	1035	1378	1924
1.15	199.78	138.74	990	1318	1842
1.175	204.12	141.75	949	1263	1764
1.2	208.47	144.77	910	1211	1691
1.225	212.81	147.78	873	1162	1623
1.25	**217.15**	**150.8**	**838**	**1116**	**1559**
1.275	221.5	153.82	806	1073	1498
1.3	225.84	156.83	775	1032	1441
1.325	230.18	159.85	746	993	1387
1.35	234.52	162.86	719	957	1336
1.375	238.87	165.88	693	922	1288
1.4	243.21	168.9	668	890	1243
1.425	247.55	171.91	645	859	1199
1.45	251.9	174.93	623	829	1158
1.475	256.24	177.94	602	801	1119
1.5	**260.58**	**180.96**	**582**	**775**	**1082**

TABLE : G26 Velocity, Discharge and Gradients for different Values of Mannings 'n' for Gravity Flow with Pipe Running Full.

Diameter: 1700 mm **Area : 2.2698 sqm**

Velocity	*Flow*		*Gradient : 1 : L*		
V, m/s	*ML/d*	*kL/min*	*n = 0.015*	*n = 0.013*	*n = 0.011*
1	*2*	*3*	*4*	*5*	*6*
0.4	78.45	54.48	8876	11815	16503
0.425	83.35	57.88	7863	10466	14619
0.45	88.25	61.28	7013	9336	13039
0.475	93.15	64.69	6295	8379	11703
0.5	**98.06**	**68.1**	**5681**	**7562**	**10562**
0.525	102.96	71.5	5153	6859	9580
0.55	107.86	74.9	4695	6249	8729
0.575	112.77	78.31	4296	5718	7986
0.6	117.67	81.72	3945	5251	7335
0.625	122.57	85.12	3636	4840	6760
0.65	127.48	88.53	3361	4474	6250
0.675	132.38	91.93	3117	4149	5795
0.7	137.28	95.33	2898	3858	5389
0.725	142.18	98.74	2702	3597	5024
0.75	**147.09**	**102.15**	**2525**	**3361**	**4694**
0.775	151.99	105.55	2365	3148	4396
0.8	156.89	108.95	2219	2954	4126
0.825	161.8	112.36	2087	2778	3880
0.85	166.7	115.76	1966	2617	3655
0.875	171.6	119.17	1855	2469	3449
0.9	176.5	122.57	1753	2334	3260
0.925	181.41	125.98	1660	2209	3086
0.95	186.31	129.38	1574	2095	2926
0.975	191.21	132.78	1494	1989	2778
1	**196.12**	**136.19**	**1420**	**1890**	**2640**
1.025	201.02	139.6	1352	1799	2513
1.05	205.92	143	1288	1715	2395
1.075	210.82	146.4	1229	1636	2285
1.1	215.73	149.81	1174	1562	2182
1.125	220.63	153.22	1122	1494	2086
1.15	225.53	156.62	1074	1429	1997
1.175	230.44	160.03	1029	1369	1913
1.2	235.34	163.43	986	1313	1834
1.225	240.24	166.83	946	1260	1760
1.25	**245.14**	**170.24**	**909**	**1210**	**1690**
1.275	250.05	173.65	874	1163	1624
1.3	254.95	177.05	840	1119	1562
1.325	259.85	180.45	809	1077	1504
1.35	264.76	183.86	779	1037	1449
1.375	269.66	187.26	751	1000	1397
1.4	274.56	190.67	725	965	1347
1.425	279.46	194.07	699	931	1300
1.45	284.37	197.48	675	899	1256
1.475	289.27	200.88	653	869	1214
1.5	**294.17**	**204.28**	**631**	**840**	**1174**

TABLE : G27 Velocity, Discharge and Gradients for different Values of Mannings 'n' for Gravity Flow with Pipe Running Full.

Diameter: 1800 mm **Area : 2.5447 sqm**

Velocity V, m/s	Flow ML/d	Flow kL/min	Gradient : 1 : L n = 0.015	Gradient : 1 : L n = 0.013	Gradient : 1 : L n = 0.011
1	2	3	4	5	6
0.4	87.95	61.08	9579	12751	17810
0.425	93.44	64.89	8485	11295	15776
0.45	98.94	68.71	7569	10075	14072
0.475	104.44	72.53	6793	9042	12630
0.5	**109.93**	**76.34**	**6131**	**8161**	**11398**
0.525	115.43	80.16	5561	7402	10338
0.55	120.93	83.98	5067	6744	9420
0.575	126.42	87.79	4636	6171	8619
0.6	131.92	91.61	4257	5667	7915
0.625	137.42	95.43	3924	5223	7295
0.65	142.91	99.24	3628	4829	6744
0.675	148.41	103.06	3364	4478	6254
0.7	153.91	106.88	3128	4164	5815
0.725	159.4	110.69	2916	3881	5421
0.75	**164.9**	**114.51**	**2725**	**3627**	**5066**
0.775	170.4	118.33	2552	3397	4744
0.8	175.89	122.15	2395	3188	4452
0.825	181.39	125.97	2252	2997	4187
0.85	186.89	129.78	2121	2824	3944
0.875	192.38	133.6	2002	2665	3722
0.9	197.88	137.42	1892	2519	3518
0.925	203.38	141.24	1791	2384	3330
0.95	208.87	145.05	1698	2261	3157
0.975	214.37	148.87	1612	2146	2998
1	**219.87**	**152.69**	**1533**	**2040**	**2850**
1.025	225.36	156.5	1459	1942	2712
1.05	230.86	160.32	1390	1850	2585
1.075	236.36	164.14	1326	1765	2466
1.1	241.85	167.95	1267	1686	2355
1.125	247.35	171.77	1211	1612	2251
1.15	252.85	175.59	1159	1543	2155
1.175	258.34	179.4	1110	1478	2064
1.2	263.84	183.22	1064	1417	1979
1.225	269.34	187.04	1021	1360	1899
1.25	**274.83**	**190.85**	**981**	**1306**	**1824**
1.275	280.33	194.67	943	1255	1753
1.3	285.83	198.49	907	1207	1686
1.325	291.32	202.31	873	1162	1623
1.35	296.82	206.13	841	1119	1564
1.375	302.32	209.94	811	1079	1507
1.4	307.81	213.76	782	1041	1454
1.425	313.31	217.58	755	1005	1403
1.45	318.81	221.4	729	970	1355
1.475	324.3	225.21	704	938	1310
1.5	**329.8**	**229.03**	**681**	**907**	**1266**

TABLE : G28 Velocity, Discharge and Gradients for different Values of Mannings 'n' for Gravity Flow with Pipe Running Full.

Diameter: 1900 mm **Area : 2.8353 sqm**

Velocity V, m/s	*Flow*		*Gradient : 1 : L*		
	ML/d	*kL/min*	*n = 0.015*	*n = 0.013*	*n = 0.011*
1	2	3	4	5	6
0.4	97.99	68.05	10295	13704	19141
0.425	104.11	72.3	9120	12139	16955
0.45	110.24	76.56	8134	10828	15123
0.475	116.36	80.81	7301	9718	13573
0.5	**122.49**	**85.06**	**6589**	**8770**	**12250**
0.525	128.61	89.31	5976	7955	11111
0.55	134.74	93.57	5445	7248	10124
0.575	140.86	97.82	4982	6632	9263
0.6	146.98	102.07	4576	6091	8507
0.625	153.11	106.33	4217	5613	7840
0.65	159.23	110.58	3899	5190	7249
0.675	165.36	114.83	3615	4812	6722
0.7	171.48	119.08	3362	4475	6250
0.725	177.61	123.34	3134	4171	5826
0.75	**183.73**	**127.59**	**2928**	**3898**	**5444**
0.775	189.86	131.85	2743	3651	5099
0.8	195.98	136.1	2574	3426	4785
0.825	202.1	140.35	2420	3221	4500
0.85	208.23	144.6	2280	3035	4239
0.875	214.35	148.85	2151	2864	4000
0.9	220.48	153.11	2034	2707	3781
0.925	226.6	157.36	1925	2563	3579
0.95	232.73	161.62	1825	2429	3393
0.975	238.85	165.87	1733	2306	3222
1	**244.97**	**170.12**	**1647**	**2193**	**3062**
1.025	251.1	174.38	1568	2087	2915
1.05	257.22	178.63	1494	1989	2778
1.075	263.35	182.88	1425	1897	2650
1.1	269.47	187.13	1361	1812	2531
1.125	275.6	191.39	1302	1732	2420
1.15	281.72	195.64	1246	1658	2316
1.175	287.85	199.9	1193	1588	2218
1.2	293.97	204.15	1144	1523	2127
1.225	300.09	208.4	1098	1461	2041
1.25	**306.22**	**212.65**	**1054**	**1403**	**1960**
1.275	312.34	216.9	1013	1349	1884
1.3	318.47	221.16	975	1297	1812
1.325	324.59	225.41	938	1249	1744
1.35	330.72	229.67	904	1203	1680
1.375	336.84	233.92	871	1160	1620
1.4	342.96	238.17	840	1119	1562
1.425	349.09	242.42	811	1080	1508
1.45	355.21	246.67	783	1043	1457
1.475	361.34	250.93	757	1008	1408
1.5	**367.46**	**255.18**	**732**	**974**	**1361**

TABLE : G29 Velocity, Discharge and Gradients for different Values of Mannings 'n' for Gravity Flow with Pipe Running Full.

Diameter: 2000 mm **Area : 3.1416 sqm**

Velocity V, m/s	Flow ML/d	Flow kL/min	Gradient : 1 : L n = 0.015	n = 0.013	n = 0.011
1	2	3	4	5	6
0.4	108.58	75.4	11024	14674	20495
0.425	115.36	80.11	9765	12998	18155
0.45	122.15	84.83	8710	11594	16194
0.475	128.93	89.53	7818	10406	14534
0.5	**135.72**	**94.25**	**7055**	**9391**	**13117**
0.525	142.51	98.97	6399	8518	11897
0.55	149.29	103.67	5831	7761	10840
0.575	156.08	108.39	5335	7101	9918
0.6	162.86	113.1	4900	6522	9109
0.625	169.65	117.81	4515	6010	8395
0.65	176.44	122.53	4175	5557	7761
0.675	183.22	127.24	3871	5153	7197
0.7	190.01	131.95	3600	4791	6692
0.725	196.79	136.66	3356	4467	6239
0.75	**203.58**	**141.38**	**3136**	**4174**	**5830**
0.775	210.37	146.09	2937	3909	5460
0.8	217.15	150.8	2756	3668	5124
0.825	223.94	155.51	2591	3449	4818
0.85	230.72	160.22	2441	3250	4539
0.875	237.51	164.94	2304	3066	4283
0.9	244.3	169.65	2178	2898	4048
0.925	251.08	174.36	2061	2744	3833
0.95	257.87	179.08	1954	2601	3633
0.975	264.65	183.78	1855	2470	3450
1	**271.44**	**188.5**	**1764**	**2348**	**3279**
1.025	278.23	193.22	1679	2235	3121
1.05	285.01	197.92	1600	2129	2974
1.075	291.8	202.64	1526	2032	2838
1.1	298.58	207.35	1458	1940	2710
1.125	305.37	212.06	1394	1855	2591
1.15	312.16	216.78	1334	1775	2480
1.175	318.94	221.49	1278	1701	2375
1.2	325.73	226.2	1225	1630	2277
1.225	332.51	230.91	1175	1565	2185
1.25	**339.3**	**235.63**	**1129**	**1503**	**2099**
1.275	346.09	240.34	1085	1444	2017
1.3	352.87	245.05	1044	1389	1940
1.325	359.66	249.76	1005	1337	1868
1.35	366.44	254.47	968	1288	1799
1.375	373.23	259.19	933	1242	1734
1.4	380.02	263.9	900	1198	1673
1.425	386.8	268.61	869	1156	1615
1.45	393.59	273.33	839	1117	1560
1.475	400.37	278.03	811	1079	1507
1.5	**407.16**	**282.75**	**784**	**1043**	**1457**

SECTION-III

Hydraulic Parameters of Pipes Flowing Partially Full

TABLE : PF01 Hydraulic Parameters of Partially Flowing Pipes, 'n' constant

$dr = d/D$	*Angle*	$pr = p/P$	$ar = a/A$	$rr = r/R$	$vr = v/V$	$qr = q/Q$
1	2	3	4	5	6	7
0.1	1.287	0.2048	0.052	0.2541	0.4012	0.0209
0.2	1.8546	0.2952	0.1424	0.4824	0.6151	0.0876
0.3	2.3186	0.369	0.2523	0.6838	0.7762	0.1958
0.4	2.7389	0.4359	0.3735	0.8569	0.9022	0.337
0.5	3.1416	0.5	0.5	1	1	0.5
0.6	3.5443	0.5641	0.6265	1.1106	1.0724	0.6719
0.7	3.9646	0.6315	0.7477	1.1849	1.1198	0.8373
0.8	4.4286	0.7048	0.8576	1.2168	1.1398	0.9775
0.9	4.9962	0.7952	0.948	1.1921	1.1243	1.0658
1	6.2832	1	1	1	1	1

Notes:

1. Angle is given in radian (1 radian= 57.295 degree),
2. Small letters refer to partilly flowing section and capital letters refer to fully flowing section.

TABLE : PF02 Hydraulic Parameters of Partially Flowing Pipes, 'n' constant

$dr = d/D$	Angle	$pr = p/P$	$ar = a/A$	$rr = r/R$	$vr = v/V$	$qr = q/Q$
1	2	3	4	5	6	7
0.1	1.287	0.2048	0.052	0.2541	0.4012	0.0209
0.12	1.415	0.2252	0.068	0.3018	0.4499	0.0306
0.14	1.534	0.2441	0.0851	0.3486	0.4953	0.0422
0.16	1.6461	0.262	0.1033	0.3942	0.5376	0.0555
0.18	1.7526	0.2789	0.1224	0.4388	0.5774	0.0707
0.2	1.8546	0.2952	0.1424	0.4824	0.6151	0.0876
0.22	1.9528	0.3108	0.1631	0.5248	0.6506	0.1061
0.24	2.0479	0.3259	0.1846	0.5662	0.6844	0.1263
0.26	2.1403	0.3406	0.2066	0.6065	0.7165	0.148
0.28	2.2304	0.355	0.2292	0.6457	0.747	0.1712
0.3	2.3186	0.369	0.2523	0.6838	0.7762	0.1958
0.32	2.4051	0.3828	0.2759	0.7207	0.8038	0.2218
0.34	2.4901	0.3963	0.2998	0.7565	0.8302	0.2489
0.36	2.574	0.4097	0.3241	0.7911	0.8554	0.2772
0.38	2.6569	0.4229	0.3487	0.8246	0.8793	0.3066
0.4	2.7389	0.4359	0.3735	0.8569	0.9022	0.337
0.42	2.8202	0.4488	0.3986	0.888	0.9239	0.3683
0.44	2.901	0.4617	0.4238	0.9179	0.9445	0.4003
0.46	2.9814	0.4745	0.4491	0.9465	0.964	0.4329
0.48	3.0616	0.4873	0.4746	0.9739	0.9825	0.4663
0.5	3.1416	0.5	0.5	1	1	0.5
0.52	3.2216	0.5127	0.5255	1.0248	1.0165	0.5342
0.54	3.3018	0.5255	0.5509	1.0483	1.0319	0.5685
0.56	3.3822	0.5383	0.5762	1.0705	1.0465	0.603
0.58	3.463	0.5512	0.6014	1.0912	1.0599	0.6374
0.6	3.5443	0.5641	0.6265	1.1106	1.0724	0.6719
0.62	3.6263	0.5771	0.6513	1.1285	1.0839	0.7059
0.64	3.7092	0.5903	0.6759	1.1449	1.0944	0.7397
0.66	3.7931	0.6037	0.7002	1.1599	1.1039	0.773
0.68	3.8781	0.6172	0.7241	1.1732	1.1124	0.8055
0.7	3.9646	0.631	0.7477	1.1849	1.1198	0.8373
0.72	4.0528	0.645	0.7708	1.195	1.1261	0.868
0.74	4.1429	0.6594	0.7934	1.2033	1.1313	0.8976
0.76	4.2353	0.6741	0.8154	1.2097	1.1353	0.9257
0.78	4.3304	0.6892	0.8369	1.2143	1.1382	0.9526
0.8	4.4286	0.7048	0.8576	1.2168	1.1398	0.9775
0.82	4.5306	0.7211	0.8776	1.2171	1.14	1.0005
0.84	4.6371	0.738	0.8967	1.215	1.1386	1.021
0.86	4.7492	0.7559	0.9149	1.2104	1.1358	1.0391
0.88	4.8682	0.7748	0.932	1.2029	1.1311	1.0542
0.9	4.9962	0.7952	0.948	1.1921	1.1243	1.0658
0.92	5.1362	0.8174	0.9625	1.1775	1.1151	1.0733
0.94	5.2933	0.8425	0.9755	1.1579	1.1027	1.0757
0.96	5.4778	0.8718	0.9866	1.1316	1.0859	1.0713
0.98	5.7156	0.9097	0.9952	1.0941	1.0618	1.0567
1	6.2832	1	1	1	1	1

Notes: 1. Angle is given in radian.(1 radian = 57.295 degree)

2. Small letters refer to partially flowing section of pipe and capital letters refer to fully flowing section of pipe.

TABLE : PF03 Hydraulic Parameters of Partially Flowing Pipes, '*n*' variable

dr = *d*/*D*	*Angle*	*nr* = *n*/*N*	*ar* = *a*/*A*	*rr* = *r*/*R*	*vr* = *v*/*V*	*qr* = *q*/*Q*
1	2	3	4	5	6	7
0.1	1.287	1.22	0.052	0.2541	0.3288	0.0171
0.2	1.8546	1.27	0.1424	0.4824	0.4843	0.069
0.3	2.3186	1.28	0.2523	0.6838	0.6064	0.153
0.4	2.7389	1.27	0.3735	0.8569	0.7104	0.2653
0.5	3.1416	1.24	0.5	1	0.8065	0.4033
0.6	3.5443	1.21	0.6265	1.1106	0.8863	0.5553
0.7	3.9646	1.18	0.7477	1.1849	0.9489	0.7095
0.8	4.4286	1.14	0.8576	1.2168	0.9998	0.8574
0.9	4.9962	1.07	0.948	1.1921	1.0507	0.9961
1	6.2832	1	1	1	1	1

Notes : 1. Angle is given in radian. (1 radian = 57.295 degree)

2. Small letters refer to partilly flowing section of pipe and capital letters refer to fully flowing section of pipe